FEARVANA

Advance Praise for *Fearvana*

"Akshay Nanavati will change the way you think about fear and by doing so change your life for the better."

—**Cal Newport**, author of *Deep Work*

"Out of the ashes of combat and trauma, Akshay Nanavati has written a gripping, hopeful, helpful book combining cutting-edge neuroscience, solid tools from mindfulness and psychology, and his own hard-won personal wisdom. A treat and a gem."

—**Rick Hanson**, PhD, author of *Hardwiring Happiness: The New Brain Science of Contentment, Calm, and Confidence*

"Fear is an emotional response to risk that often leads to self-doubt and despondency. Akshay Nanavati's *Fearvana* is an effective personal guide to understand and overcome the many aspects of fear psychosis—a must-read for all risk takers in this new age of start-ups where stress-related failure is inherent."

—**Kiran Mazumdar-Shaw**, biotech entrepreneur and one of *Forbes'* 100 most powerful women in the world

"No matter how bad your life seems, you can still experience bliss. *Fearvana* shows you how."

—**Charlie Hoehn**, author of *Play It Away: A Workaholic's Cure for Anxiety*

"This is not the standard personal development book you are used to. Every chapter is packed with a wealth of inspiring stories, cutting-edge insights, and practical takeaways that finally reveal the secrets of what it really takes to change behavior. *Fearvana* is, without a doubt, one of the few books that really does stand out as a must-read book."

—**Jack Canfield**, coauthor of the #1 *New York Times*–bestselling Chicken Soup for the Soul® series, *Dare to Win*, and *The Success Principles*™

"Akshay Nanavati writes with power and conviction. He is a man who has been through more struggle and suffering than most of us will ever know. Yet through it all he has remained resolute, developing a remarkable resiliency.

The lessons he imparts in *Fearvana* are timeless wisdom that serve to better all of mankind."

—**Dean Karnazes**, ultramarathoner and *New York Times*–bestselling author

"This book is unlike anything you have ever read before. It will inspire you, challenge you, educate you, and it might even shock you, but one thing is for sure, it will transform you. Do yourself a favor and pick up this book today!"

—**Steve Olsher**, *New York Times* bestselling author of *What Is Your WHAT?* and host of the #1-rated radio show/podcast Reinvention Radio

"We waste too much of our lives on fear. Akshay Nanavati offers a persuasive and evidence-based set of guidelines for taking back control and optimizing for your true goals."

—**David Rowan**, editor of *WIRED* magazine

"It's one thing to preach about turning adversity into strengths. It's another to come back from the brink of death and despair actually stronger, then build a clear, step-by-step process for anyone to understand to transform fear and anxiety. Well researched, powerful, and clearly written, this book is a goldmine for finding enduring courage and greater contentment."

—**Jaimal Yogis**, author of *The Fear Project* and *All Our Waves Are Water*

"In his extraordinary book *Fearvana*, Akshay Nanavati not only sheds light on the science of fear—what it is and why we need it—he outlines a roadmap for transforming fear into fuel. Imminently readable and highly recommended, this is a great book for anyone who wants to take life to the next level."

—**Libby Gill**, executive coach, leadership expert, and author of *You Unstuck*

"*Fearvana* is a gold mine for successful people who inevitably encounter fear, stress, and anxiety as they take on big challenges. Akshay Nanavati makes a compelling case that we have much to gain by accepting and even embracing our fears instead of shutting them out. Only then can we fulfill our potential."

—**Marshall Goldsmith**, executive coach, business educator, and *New York Times*–bestselling author, ranked the #1 leadership thinker in the world by Thinkers50

"Whether it is fitness, running a business, writing a book, or any area of life, Akshay understands the fundamental truth that the mind is primary when it comes to accomplishing goals. By applying the lessons learned in *Fearvana*, you will be able to harness the power of your mind and be able to overcome any challenge or obstacle that stands in your way."

—**Bobby Maximus**, bestselling author and Executive Director, Gym Jones

"*Fearvana* isn't just tips and exercises to help people overcome their fears. It's a journey through the human mind, through Akshay's mind, that gets at the heart of why we fear and how to use it to reach our highest potential. Deeply personal in its storytelling, *Fearvana* is ultimately a hopeful book about embracing and listening to our fears so we can find our paths forward."

—**Keith Ferrazzi**, #1 *New York Times*–bestselling author of *Who's Got Your Back* and *Never Eat Alone*

"Akshay Nanavati has written an exciting book that reshapes your greatest fears into love and meaning. You are stronger than you think—find out how in *Fearvana*."

—**Deborah Sandella**, PhD, RN, #1 international bestselling author of *Goodbye, Hurt & Pain: 7 Simple Steps to Health, Love, and Success*

"Life is meant to be lived beyond the edge of our comfort zone. That is where we find happiness, evolve into our greatest self, and come alive. In a profound and masterful way, *Fearvana* will teach you how to uncover the hidden reserves of courage that you never knew you had. This is your manual for thriving in the unknown."

—**Sebastian Copeland**, award-winning photographer and one of the world's top 50 adventurers

"A deeply personal journey of struggles, survival, and ultimate triumph through harnessing the power of fear! *Fearvana* provides a very insightful roadmap on how to gain back self-identity and personal happiness by conquering fear, anxiety, and stress. It is destined to become a bestseller!"

—**Jagdish N. Sheth**, Professor of Business, Emory University

"Counterintuitive, practical, and potentially life-changing, Akshay's book wants to rewire the way you look at fear."

—**Seth Godin**, author of *Linchpin*

"If your fear, stress, or anxiety are standing in the way of having the life you want, then read this powerful book. It will show you how to make those feelings work for you rather than against you. Filled with inspiring stories, in-depth research, and actionable takeaways, *Fearvana* will guide you to create a life of greater happiness and success."

—**Marci Shimoff**, #1 *New York Times* bestselling author of
Happy for No Reason and *Chicken Soup for the Woman's Soul*

"Who better to teach us about pushing our limits in a positive way than someone who has fought in Iraq, climbed the Himalayas, and crossed the ice cap in Greenland? Akshay writes about challenging topics in a way that is easy to understand, and demonstrates that the beauty of life is that we all have room to grow. As he so clearly states, we should celebrate and acknowledge ourselves for our hard work, not just our results—I am going to incorporate that into my daily life. Thank you, Akshay, for your deep introspection and meaningful guidance."

—**Justin Constantine**, inspirational speaker and
Lieutenant Colonel, U.S. Marine Corps (Ret.)

FEARVANA

The Revolutionary Science of How
to Turn Fear into Health, Wealth and Happiness

AKSHAY NANAVATI

NEW YORK

NASHVILLE • MELBOURNE • VANCOUVER

FEARVANA
The Revolutionary Science of How to
Turn Fear into Health, Wealth and Happiness

Published in New York, New York, by Morgan James Publishing. Morgan James is a trademark of Morgan James, LLC. www.MorganJamesPublishing.com

The Morgan James Speakers Group can bring authors to your live event. For more information or to book an event visit The Morgan James Speakers Group at www.TheMorganJamesSpeakersGroup.com.

ISBN 978-1-63047-605-2 paperback
ISBN 978-1-63047-606-9 eBook
Library of Congress Control Number: 2015917592

Cover Design by:
Chris Treccani
www.3dogdesign.net

Interior Design by:
Bonnie Bushman
The Whole Caboodle Graphic Design

In an effort to support local communities, raise awareness and funds, Morgan James Publishing donates a percentage of all book sales for the life of each book to Habitat for Humanity Peninsula and Greater Williamsburg.

Get involved today! Visit
www.MorganJamesBuilds.com

To my global human family

TABLE OF CONTENTS

FOREWORD

We all cherish the precious gift of human life. However, as we strive to make the best of it, we often face experiences that cause us fear and hopelessness. Many of our problems are self-inflicted, resulting from our lack of concern for the wellbeing of others and from our extreme self-centeredness. When confronted with challenges that might affect our spirit, it is helpful to view the situation constructively; our own hardships can actually boost our determination and instill in us a sense of concern for those facing similar difficulties.

I am happy that Akshay Nanavati has found the strength to overcome the traumatic experiences of his life and has developed the aspiration to help others. His book, *Fearvana*, inspires us to look beyond our own agonizing experiences, suggesting means for overcoming our fears. I appreciate his sincerity and hope that others will find reason and the encouragement to see the positive side of their lives.

His Holiness the Dalai Lama

Introduction

FROM THE BRINK OF SUICIDE
TO THE SCIENCE OF FEARVANA

"The graveyard is the richest place on earth, because it is here that you will find all the hopes and dreams that were never fulfilled, all because someone was too afraid to take that first step, keep with the problem, or determined to carry out their dream."

—Les Brown

I lay sprawled out on the sofa, gazing into the barren expanse above. My arm dangled off the edge. Below it rested an empty, one-liter bottle of vodka. The sound of a TV echoing in the background shielded me from an army of demons lingering at the gates. They waited for the inevitable: a single moment of stillness to storm into the emptiness of my soul.

The sun's rays pierced through the pale, white shutters, disrupting my lifeless slumber. Awakened from my daze, I mustered a herculean effort to wrestle my feet onto the ground. Hunching over with both hands pressing against my skull, I saw the bottle of vodka glaring at me from the abyss—an excruciating reminder of the depths to which I had fallen.

As I looked up to get a glimpse of the outside world, the light stole my attention away from the noise, robbing me of my only comfort. A deafening roar of silence descended upon me. Exposed and vulnerable to the chaos of my consciousness, a week of nonstop binge drinking finally took its toll. *This pattern will never change*, I thought to myself. *I will never change.*

Entombed within this pit of despair, all hope faded away. What was the point of going on? For the first and last time in my life, I contemplated the unthinkable. The severity of such a thought infiltrating my mind shocked me out of my stupor. How could I even think of taking my own life? Forcing myself upright, I stared back into the light. This time, it became my ally, illuminating a way out of the darkness.

The Single Most Important Skill You Need to Succeed at Anything

"Adversity is the diamond dust Heaven polishes its jewels with."
—**Thomas Carlyle**

I am guessing you picked up this book because you too have gone through challenges in your life, and at this very moment, you still have obstacles to overcome. You have probably figured out by now that it is impossible to live without some degree of hardship.

The good news for you is that no matter what has stopped you in the past or what is holding you back right now, it is not entirely your fault. As my friend and mentor Jack Canfield once told me, we all do the best we can with the level of skills, abilities, knowledge, and awareness we possess at any given time.

Allow this to really sink in. You can relax and stop beating yourself up. Any time you might be feeling confused, lost, or scared, it is not your fault. Section 1 will show you why that is and how accepting this comforting fact is the first stepping stone to transforming your life.

My goal is for you to finish this book with a new level of awareness and a clear set of actions that will enable you to achieve mastery over your most valuable asset: your mind. The mind is the machine that controls everything in your world. How you respond to any event and how you navigate the obstacles standing in your way is all dependent on your ability to direct your mind—instead of it directing you.

When you become the director, you will have the superhuman ability to boldly face any challenge that comes your way, whether it be training for a marathon, raising a child, finding the perfect career, building your own business, recovering from a traumatic event, or figuring out your purpose in life—anything!

If you put the systems in this book to use, the event itself will become irrelevant. By the end of section 3, you will know how to transform fear, stress, and anxiety into your allies, regardless of the context that creates them.

In life, you cannot predict the next wave that will hit you on your voyage to greatness, but you can develop the habit of braving the stormy seas, no matter how many waves crash into you. This is the most important skill you could ever learn.

After decades of research, in what has become the largest and most important study on happiness, professor Mihaly Csikszentmihalyi, bestselling author of *Flow: The Psychology of Optimal Experience*, concluded, "Of all the virtues we can learn, no trait is more useful, more essential for survival, and more likely to improve the quality of life than the ability to transform adversity into an enjoyable challenge." Teaching myself this trait brought me back from the brink of death.

So how did I do it? And how can you master this ability? By the end of this book, you will know the answer to those questions and others, like these:

- How are fear, stress, and anxiety the most powerful weapons to beat fear, stress, and anxiety?
- What six simple words led to a 30 percent improvement in student test scores? (Parents, this makes all the difference in your child's future.)
- What does it really mean to be happy?
- What is the one habit no one talks about, yet it is more important than any other to cultivate?
- How can you travel through time and change your past?

These questions carry within them secrets that could transform your life forever. If you want to unlock them, stop reading and take as long as you need to answer these questions:

- What challenges are you facing right now?
- How and where is fear, stress, or anxiety holding you back?

There is no one-size-fits-all model for transformation. How you interpret the following pages will be different from everyone else, so it is essential for you to be clear on what you want to get out of this book. Clarity is power. Without it, everything you are about to read will just be more information.

I do not want this to be another book that makes no real difference for you. For this, or any other book to work for you, you must analyze, absorb, and act upon the information from the perspective of your own needs. That requires you to approach the material with an open mind, willing to seek out the value in every page and then choose whether or not to apply it to your life. As the Zen master Shunryu Suzuki said, "In the beginner's mind there are many possibilities, but in the expert's there are few." An open mind creates the space for new perspectives that lead to wisdom, and more importantly, transformation.

By asking you to hold on to a beginner's mind, I am asking you to place your trust in me. I do not take that responsibility lightly. As such, all the information in the following pages will be validated either by scientific evidence, real world examples, spiritual teachings, in-depth studies on human psychology, or all of the above. These ideas are not something I came up with overnight. I discovered them through years of digging into what it takes to master our minds and achieve success, in every sense of the word.

The desire to take on such exhaustive research sprang from all the obstacles I have faced, including drug addiction, war, adventures in the most hostile environments on the planet, and, of course, my close encounter with the Grim Reaper. Through my studies and my life experiences, I came to learn there are invaluable gifts offered by adversity. Unearthing these gifts spawned the rise of Fearvana. It is a state of being and a lifestyle needed now more than ever.

Have We Really Made Any Progress as a Species?

Today we are living longer, healthier, and safer lives than at any other time in human history. From 1990 to 2011, life expectancy across the globe has gone up from sixty-four to seventy years. In developed countries, people even get to enjoy those extra years. According to a 2013 Harvard study, medical advancements have helped people stay healthier for longer than ever before.

For the most part, life has improved in developing countries as well. This is summarized best by the United Nations Human Development Index, which ranks every country based on the combined measurements of income, health, and literacy. Almost all of the lowest countries on the list are now better off than they used to be.

Life has not only become better for our entire human family, but we are also at less risk of losing it. A major 2011 study by Harvard professor Steven Pinker concluded, "We may be living in the most peaceful era in our species' existence."

By examining the data, the studies, and the voice of the experts in their field, it is clear that life in our world today is safer, more comfortable, and easier than the one our ancestors inhabited. Yet, we are no happier or healthier than they were. One out of every three adults and children is considered to be obese. Obesity has more than doubled across the globe in the last twenty-four years. Today, one in five Americans is suffering from "extreme stress." One in every five adults and teenagers live with some sort of mental illness.

This is not just a problem for Americans. In India, over fifty million people live with mental illness. One in four people across the planet will be affected by a mental disorder at some point in his life. Depression is the leading cause of disability in the world. The World Health Organization states that three hundred and fifty million people globally suffer from it. That figure is more than the entire population of the United States. Worst of all, within the next two minutes, three people will take their own life somewhere across the globe. I could go on and on.

"Why is it that, despite having achieved previously undreamed-of miracles of progress, we seem more helpless in facing life than our less privileged ancestors were?" asked Csikszentmihalyi during his research on happiness. "The answer seems clear. While humankind collectively has increased its material powers a thousandfold, it has not advanced very far in terms of improving the content of experience."

The irony is that the progress we have made to make our lives better is, in many ways, making it worse. Psychologists David Myers and Robert Lane found that as the gross domestic product increased by more than 50 percent over the last thirty years, this led to a 5 percent decrease in happiness. Their studies have revealed greater affluence and a greater number of choices has led to more depression, loneliness, anxiety, stress, and overall decrease in well-being.

Today we have more choices, opportunities, and distractions than ever before. The more choices we have, the more confused we have become. Psychologists call it the paradox of choice. What this all means is excessive security and comfort have become our burden. "We are the healthiest, wealthiest, and longest-lived people in history. And we are increasingly afraid," writes Daniel Gardner, author of *The Science of Fear*.

Don't worry, there is a solution that doesn't involve returning to the Stone Age. Although as a whole, we might be struggling to find joy in the moment-to-moment adventure that is life, there are a select few who have found the secret to developing mastery over the mind and the self, regardless of their external reality. All I have done is put a label to it. The rest of the book will show you how anyone, including you, can implement this secret.

What Is Fearvana?

"It is not because things are difficult that we do not dare, it is because we do not dare that they are difficult."

—**Seneca**

Imagine you are standing backstage, getting ready to share your story and your message to millions of people. Your heart starts beating faster. Your palms become sweaty. The butterflies in your stomach go wild. You begin wondering, *What if people hate me? What if they laugh at me? What if they reject me?* This only makes you more nervous. Then Oprah announces your name. You pause.

You breathe it all in. A few seconds later, you run out onto the stage in front of the world.

Now imagine yourself sitting at home on your couch. Within you lies an idea for a business or a book. Deep down, you know this is a great idea that could transform a countless number of lives. If you act on it, it could change your life as well. You would have more money and more freedom to do whatever you wanted, whenever you wanted. You start to get excited. Then you start to wonder, *What if I fail? What if no one likes what I create? What if I can't do it?* Once again, you feel the fear, stress, and anxiety. But only this time, you let them win. The former experience is Fearvana. The latter experience is the real suffering that occurs when you choose not to engage your fears—the same kind of suffering I experienced and described at the beginning of this book.

If you don't choose a worthy struggle, struggle will choose you. And when it does, it will do far worse damage than the pain you will inevitably feel on the journey to greatness. As entrepreneur and bestselling author Jim Rohn said, "We must all suffer from one of two pains: the pain of discipline or the pain of regret. The difference is discipline weighs ounces while regret weighs tons."

Pain is not a bad thing. The greater your struggle, the greater the rewards from the victory on the other side. Which is why we, as a society, revere those who have triumphed over adversity to achieve success.

Think about it. Whom do you admire most in the world? When Csikszentmihaly asked that question to people all over the world, he found "courage and the ability to overcome hardship are the qualities most often mentioned as a reason for admiration." Through Fearvana, you too can develop those qualities in any situation.

Fearvana is the bliss that results from engaging our fears to pursue our own worthy struggle. Fearvana is running a marathon, building a business, writing a book, or anything meaningful you pursue in service of your growth and happiness. At that intersection of pain and pleasure, we become the greatest version of ourselves.

Neurologically, these two seemingly contradictory forces are two sides of the same coin. "There's a substantial overlap between those brain areas involved in processing fear and pleasure," said Allan Kalueff, a neuroscientist at the University of Tampere in Finland. Your brain is wired to connect fear and bliss into one, all-powerful force that gives you the strength to accomplish anything. When you master the art of seeking your own worthy struggle to experience bliss, you will be able to find bliss in the face of struggle, no matter how it shows up.

Wherever you are in your life at this moment, whatever challenges you might be facing, however audacious and grand your goals, embracing the Fearvana mindset and lifestyle is the secret that will get you from where you are now to where you want to be. "To really achieve anything, you have to be able to tolerate and enjoy risk. You're just never going to have a breakthrough without it," says neuropsychologist Barbara Sahakian.

If you want to achieve something you have never accomplished before or be someone you have never been before, you have to do something you have never done before. Risk is an inherent part of growth. When you take risks, your brain responds with fear. It is the first response to external stimuli because the primary job of your brain is to determine whether an unknown stimulus will kill you. Inevitably then, we need fear to succeed. Yet, fear has a bad reputation. We often hear people say things like, "Don't be scared," or "There's nothing to be afraid of," as if fear is somehow a bad thing.

I created the term *Fearvana* as a direct response to our mistaken beliefs about the value of fear and its counterparts, stress and anxiety. Like the concept of yin and yang, the concept of Fearvana illustrates how two seemingly opposing forces at their extremes, fear and nirvana, are, in fact, complementary. Fear is generally believed to be negative. Nirvana is considered the epitome of positivity—bliss and enlightenment itself. When we bring them together with focused action, we become the greatest version of ourselves.

Fearvana is a new word to transform our relationship with fear so we can harness its power. "Simply by changing your habitual vocabulary—the words you consistently use to describe the emotions in your life—you can instantly change how you think, feel, and how you live," states Tony Robbins.

Throughout these pages, I will give you plenty of examples of how Fearvana has helped people get the results they want. I will also show you the science of how Fearvana radically improves our performance and the content of our experience. I will teach you exactly how you can leverage it so you know how to use your fear, stress, and anxiety to transform into the strongest, happiest, and greatest version of yourself. When you know how to harness the gift that lies in adversity, nothing will ever again stop you from creating the life you want to live.

How to Use This Book

> *"No problem can be solved from the same level of consciousness that created it."*
>
> **—Albert Einstein**

This book is divided into three sections that outline the steps of the Fearvana system, with the end goal being mastery over the self to create optimal health, wealth, and happiness.

The first step is awareness and acceptance. This section will teach you how your brain runs your life without your knowledge and why it is not your fault when you find yourself unable to take the necessary actions to get the results you want. This section is an essential first step in alleviating the negative effects of anything that might cause you psychological anguish, including fear, stress, anxiety, depression, or PTSD. Section one lays the foundation for how to create lasting change.

Step two is taking action. This section will teach you everything you need to know about Fearvana. It will show you how to transform all your negative mental forces into your most powerful allies. When harnessed, those same forces become rocket fuel to propel you into your greatness. Section 2 will also show you how to create your individual path to becoming the person you want to be.

The final step is your awakening. This section will show you how to embrace the struggle inherent in achieving anything worthwhile and teach you how to have fun with that struggle so you can keep seeking it out. This step is about creating alignment within yourself, your environment, and the world around you to harness Fearvana as a force of good in your own life and the lives of our greater human family. What you are about to learn will not always be easy to implement. However, if you do put the lessons into practice, I promise you it will be worth the effort.

Relating to fear, stress, anxiety, and struggle as positive forces in our lives seems to go against conventional wisdom. There is a good chance what you are about to read will challenge many of your current beliefs about yourself and the world around you. But what got you here will not get you there, wherever "there" might be for you. It will require letting go of your current self to embrace the new self you want to become.

I assume you picked up this book to learn something new, better yourself in some way, and be more, have more, and/or do more in general. So if you want this book to help you, when you come up against an idea that confronts you, pause and ask yourself, "How can this information be valuable to me? What can I learn from this?" If you approach the material with a beginner's mind, you will find an answer to those questions.

However, I am not asking you to accept everything you read. I cite many studies, validate my arguments with research, and share stories to draw out my own conclusions, but feel free to make of this material whatever you want. Although I will provide action steps, I also encourage you to use your creativity to find your own ways to implement what you learn. As long as you are open to finding the value in it, as opposed to resisting it, the content in these pages will help you, even if your conclusions are different than mine.

I will show you why the material works, but rest assured this is not an academic textbook. The science will only serve to increase awareness and guide you into action. For that reason, I do not use footnotes. We have plenty of distractions to occupy space in our consciousness. My intention is not to add more clutter that might take your focus off the content. If you are interested in my research, all of the material I used in the writing of this book can be found at www.fearvana.com/resources.

Many of the stories I have chosen to share with you reflect the extremes of the human experience in order to illustrate our collective ability to overcome any and every obstacle. This presupposes a mindset which will serve you throughout this book: if one person can do something, anyone can do it. Although the stories are all true, I have altered some names to protect identities. You may not know their real names, but you can use their courage as a reference for the possibilities available to you in your own life.

The structure of the book is similar to the human mind, a complex machine that does not function in a linear manner, but rather flows back and forth. Think of it like the cycle of waves crashing onto the shore and receding back into the ocean. Similarly, one chapter will introduce an idea or a story, let it go, and the idea will once again present itself in a different context later in the book.

As you will learn in chapter 5, your memory works as a puzzle piece assembler, so I have created a set of puzzle pieces that build on each other. By the end of the book, the pieces will come together to form a complete picture. We are laying bricks one at a time to build an entire house.

This structure dictates the necessity of some degree of repetition. This is intentional, to facilitate the learning process. The critical element for your learning will be the training exercises at the end of most chapters. These exercises will translate the material into actionable steps so you can implement the material in your daily life. However, not every chapter will be accompanied with a training exercise. Sometimes the material itself will serve as the action step by increasing your awareness and creating a paradigm shift in your consciousness.

As you read this book, I want you to know you are not alone in your journey. I am with you, which is why I will be sharing many of my own stories. I will tell you some of the darkest moments of my life, as you have just read, and some of the lighter ones. I hope that by opening up my entire life to you—the good, the bad, and the ugly—you will build some level of trust in me.

We are on this dangerous and exciting voyage together, so if I can ever be of service, please feel free to reach out to me at www.fearvana.com/contact. I am committed to your success and would be honored to support you in any way I can. Unity is a vital element of the Fearvana experience, which is why you will see me use the word *we* often throughout this book.

We are in this together, but only you can take action. It is up to you to put this information into practice and keep putting one foot in front of the other to get to where you want to go. As long as you do so with commitment and focus, you will see results. But the journey to mastery does not end; as you go through the highs and the lows of life, what matters is progress, not perfection. That is the beauty of unleashing the human potential; there is always more room to grow. Fearvana is about the journey, not the destination.

On a final note, a lot of what we will be discussing is serious and intense material, so let's not forget to pause throughout and have fun with the process. Think of the training exercises as a game and make them playful. Approach this book as one big experiment. Let the games begin.

Section 1

AWARENESS AND ACCEPTANCE

"The first step toward change is awareness. The second step is acceptance."
—Nathaniel Branden

Every human creation started as a thought. From the computer you use to the plane you fly in, all of it began as an idea in someone's head. The mind has the ability to build worlds and destroy them.

This section will lay the foundation for how this incredible machine works. Just as you cannot fix a car without knowledge of how the engine works, you cannot change your life unless you know how your mind works. Don't worry; this will not be an anatomy lesson. It will simply give you what you need to know to leverage your most valuable asset.

The first step in the Fearvana system is becoming aware of the true nature of this machine and then accepting it without resistance. The more you fight against the way your mind works, the less you will be able to make it work for you.

Imagine this entire journey to mastery over the self as climbing a mountain. To make the climb, you need to learn the art of mountaineering. You must then accept the mountain is in control. At any time, as many brave mountaineers have learned, the mountain can unleash a hellish storm and leave us mere mortals helpless in the face of its fury. You cannot fight the mountain, but simply accept whatever it throws your way. This the first step in the training to reach our summits.

Chapter 1

MAKING FEAR YOUR FRIEND

"What we fear doing most is usually what we most need to do."
—Tim Ferriss

Perched behind an M240 machine gun, the gunner opened fire, showering nine hundred and fifty rounds a minute at the speeding vehicle in a desperate attempt to save his brothers.

Two men in a pitch-black sedan raced toward the squad of Marines, doing everything they could to avoid the hail of gunfire raining down on them. They swerved across the road. They crouched beneath the windshield. Yet, they did not slow down.

BOOM!

They collided with one of the Humvees. The force of the impact launched three Marines thirty feet away from the havoc. Those were the lucky ones.

Two vehicles, now entangled in a fiery wreckage, spun around in circles, spraying fuel in every direction, a merry-go-round of mayhem. Flames engulfed everything: the messengers of death, the mangled metal frame of the Humvee, and the two Marines still inside it.

From less than a hundred yards away, the gunner witnessed the entire collision, but could do nothing to prevent it. In one single, life-altering moment when, as he said, "Life lost all its innocence," Lance Corporal Dale stood powerless against the dark hands of destiny. Despite the inevitability of the disaster, he now had a choice to make.

He chose to act. Dale leapt from his turret and moved cautiously toward the burning wreck, expecting a second explosion from a possible bomb on board the black sedan. Two charred corpses entombed in the fiery destruction sent a clear signal to Dale. They were no longer a threat.

Dale moved on to the Humvee. By the time he arrived, the other Marines had pulled the passenger out of the vehicle. Only the driver remained. Dale and his buddy Peter knew what needed to be done. They ran into the fire. With every effort, they were beaten back by the venomous smoke poisoning their lungs, piping hot metal burning their hands, and blazing heat scorching their bodies.

The heat then ignited the live ammunition in the vehicle. The rounds exploded with ferocity, bouncing onto every surface obstructing their path. They still kept fighting. After what felt like hours, Dale and Peter managed to pry open the door and get the driver out.

Through the blood and smoke stains on the driver's face, Dale recognized his twenty-three-year-old friend Greg. He remembered speaking to him just a few hours earlier. But this was no time for emotion to cloud action. Dale began the ABCs on Greg. He checked the Airway. It was filled with blood. He looked for signs of Breathing. He saw none. He searched for a pulse to determine blood Circulation. He found nothing.

While the other Marines called in a medical evacuation, Dale initiated CPR on Greg. There was still hope. Spitting Greg's blood out of his own mouth, Dale continued mouth-to-mouth resuscitation while Peter followed his lead with chest compressions.

Life started flowing through Greg again. His skin color indicated oxygen was entering his body. A faint pulse returned. He even attempted to breathe. Determined to keep Greg alive, Dale persevered in his efforts until the UH-64 helicopter arrived at the landing zone. Along with a handful of Marines, Dale grabbed the stretcher and ran toward "the bird."

Instinctively, they leaned over Greg to protect him from the dust and debris whipped up by the helicopter blades. Within minutes, they got him and the other wounded Marines on board. The loud roar echoed into the Iraqi desert as the chopper flew to what they all hoped would be salvation.

Dale's body and mind finally settled. "I was lost in shock. There was no sound. Everything was muffled, like I was underwater. I knelt down in the middle of it all, and my body started trembling," he wrote in his memoir *The Green Marine: An Irishman's War in Iraq.* "I tried to focus on praying and looked off into the desert because it was only there, where there was nothing, that I could find any solace."

As the rest of the Marines cleared the scene, a report came over the radio, announcing Greg was still alive. They clung to a glimmer of hope. With a sense of calm restored to his environment, Dale washed the blood off his face. It had soaked into his skin and crusted behind his fingernails.

Half an hour later, another report came in over the radio. Greg did not make it.

Fear: The Foundation from Which Life Springs Forth

"Survival prospects are poor for an animal that is not suspicious of novelty."
—**Dr. Daniel Kahneman**

Evolution has blessed you with not one but two brains that work together to shape your experience of life: the animal brain and the human brain. At the base of the animal brain is your brainstem. Like you, all animals have the ability to address basic survival needs, such as eating, drinking, reproducing, regulating the essential functions of the body, and defending themselves from external threats. The brainstem's sole purpose is to keep us alive and thriving as a species. The needs it addresses are instinctual and require no conscious processing.

The brainstem is the doorway between you and the rest of the world. It receives input in the form of electrical signals from your all your senses and sends that information to the rest of the animal brain, otherwise known as the limbic system. This part of your brain is responsible for your emotions and the habitual patterns that run your life. The animal brain uses everything it has learned from your past to act on its own without your awareness. It wouldn't make much sense for us to have to stop and consciously process the act of walking or brushing our teeth every morning, would it?

To optimize functionality, your animal brain has a gate that acts like a guard against external stimuli. This guard stands on duty 24/7 and filters out irrelevant stimuli from entering your brain so you can focus on what matters at any given moment, like the words on this page instead of the sensation of your toes. Unless you choose to activate it, this bodyguard operates from outside your conscious chain of command.

Conscious thought gives you the power to seize control of the guard and, in turn, sculpt your entire animal brain. Consciousness is the ability to process our experience of life, which means we can step outside of ourselves to think about the act of thinking. This enhanced intelligence lives in the human brain, otherwise known as the prefrontal cortex. Based on your survival needs, the guard decides what stimuli to allow into your human brain. If something is perceived as a threat, this bodyguard sends you into "fight-or-flight" mode. If not, it is allowed to move into your higher consciousness for further processing.

As Dr. Joseph LeDoux, from the Center for the Neuroscience of Fear and Anxiety at NYU, says, "Fear is the most primitive and basic emotion. It occurs when we encounter danger. An animal can put off the good stuff—eating, drinking, sex—for days. But responding to danger must be immediate, or there will be no more eating, drinking, or sex."

Survival takes the top spot on our brain's priority list. Fear is the fuel that allows our brain to execute its most important function and rapidly respond to threats. To build a positive relationship with fear, we must first accept it is a natural, human response to the unknown that occurs beyond our conscious control.

Whenever you take a risk or experience anything new, your brain asks itself, is this dangerous? It has no previous reference to compare the new experience to, so it doesn't know if this is something that will kill you. "Is this good or is this bad? This is the most basic question the limbic area addresses," says Dr. Daniel Siegel, a psychiatry professor at UCLA. Based on the answer, your body and mind react.

Dale's brain knew everything happening around him warranted fear, so it sent him into "fight-or-flight" mode. "First thing you recognize is that this situation is dangerous. A normal person should flee. Flight takes over. But in a nanosecond, you realize that this is a silly thing to do," Dale said of his experience.

Let's see how this played out inside his brain. His thalamus, the receiver of all external stimuli, deemed the situation to be life threatening. This kicked the rest of his brain into survival mode, shutting down certain parts of his rational, human brain that were no longer needed to keep him alive in that moment.

With clarity on where to devote its energy, his brain sent the information coming in from his senses to his amygdala, the fear center of the brain, which stores fear memories and determines threats, along with the hypothalamus, which activates the "fight-or-flight" response. They started pumping out an invaluable chemical cocktail of adrenaline, endorphins, testosterone, dopamine, oxytocin, and cortisol. This narrowed his attention, giving him the ability to filter everything else out and focus on the task at hand. The cocktail also ensured his metabolism went into high alert so he could handle the danger in front of him. As a result, Dale leapt from the turret before he was even consciously aware of it.

The presence of fear created the necessary conditions that allowed him to face it. Fear became his ally. Without it, rationality would have kept him imprisoned in inaction. As UCLA neurobiologist Michael Fanselow says, fear is "far, far more powerful than reason. It evolved as a mechanism to protect us from life-threatening situations, and from an evolutionary standpoint there's nothing more important than that."

Since our brain always chooses fear—ahead of reason—to keep us alive, evolutionarily speaking, it would only make sense that the brain finds a way to harness that fear. Otherwise, why would it show up in the first place? I have always wondered how fear could possibly be considered evil when it clearly is our best friend. It prevents us from dying!

Such an important job requires an outstanding skill set. Our brain is more than up to the task. It can thrive in the presence of fear and transform us into unstoppable warriors who never retreat from a fight. We just have to condition it for battle. Fear gave Dale courage, but during that same incident, it paralyzed another man from taking action. Fear can help us or hinder us, depending on how we perceive it, how we train in it, and who we choose to be outside of it.

Dale was a Marine at heart. The generations that came before him served in the military. He grew up with the desire to fight for something greater than himself. His self-identity was forged by the spirit of a warrior, so when fear found him, fight took over. Dale had training as a Marine and an EMT. This gave his animal brain references to recognize patterns in his environment and respond to them accordingly. Through practice and an unwavering belief in himself, he trained his brain to subconsciously choose to fight in the face of fear.

But fight was not the only force at play here. I put "fight-or-flight" in quotes, because it is a common misconception that those are the only two responses to fear or stress. In reality, we have the power to choose from a variety of options.

When the Humvee exploded, the emotional connection that bonded Dale with his fellow Marine released "the love hormone," oxytocin, into his brain. He recognized there was something more important than his own survival. His animal brain heard this message and directed him to run into the burning vehicle.

Along with fight, Dale's actions included what is known as the tend-and-befriend response. This triggers a desire to protect and care for the people we love. Depending on the scenario, we can choose how we want fear or stress to help us. Our weapons of choice include options like fight, flight, freeze, tend-and-befriend, or learn-and-grow, which is when we find the lessons in our experience to build a new and improved self. Initially, choosing a response requires conscious activation, but in time, these responses can become automatic, as they did for Dale.

By inoculating his brain to the experience of heightened fear, he was able to feel it, experience it, and act in the face of it. The problem is that when it comes to fear, our brain is designed for the kind of life-or-death situation Dale encountered in Iraq, not for the developed world where survival is no longer a daily concern. "Our brains were simply not shaped by life in the world as we know it now," writes Daniel Gardner in *The Science of Fear*. "They are the creation of the Old Stone Age. And since our brains really make us what we are, the conclusion to be drawn from this is unavoidable and a little unsettling. We are cavemen. Or cave-persons."

The experience of Fearvana reconnects us to this archaic world of survival our brain is designed for, but no longer living in. Today our lives have become extremely complicated, with more distractions and choices than ever before. Our brains are just not ready for such a complex world. "This ancient fear system is not perfectly adapted to modern life," echoes LeDoux. "It's being co-opted by all the non-life-threatening stresses we have in our lives, and there are a lot of them."

Without the simplicity of having to worry only about death, we now have the luxury, and burden, of many other things to worry about. Our cave-person-like brains don't know how to handle these daily concerns, so it does what it knows to do and reacts to non-life-threatening situations as if they were a saber-toothed tiger. As a result, fear lurks behind anything and everything, causing us mental anguish.

When we get jealous of someone flirting with our spouse, deep down we are afraid we will lose her. When we get angry at a child for staying out late, we are afraid something will happen to him. When we feel guilt for doing something wrong, we are either afraid we will get caught or afraid of how the action will affect our sense of self-identity. When we feel stress before a test or job interview, we are afraid we will do poorly.

At the root of every emotion holding us back, we will often find a hidden fear. It is no surprise, then, that the number one reason why people say they are not living the life they want is fear. When fear shows up inside us, it then produces an associated stress response as well. Fears often present themselves as stress, or even anxiety. Essentially the same neurological pattern occurs for all three emotions.

The trigger which activates each of them creates the subtle distinctions between the three, but they are closely related. Shifting our mindset on one of these so-called negative emotions will allow us to do the same for the others. Moving forward, whenever you read the words *fear*, *stress*, or *anxiety*, choose the one you feel most applies to your situation and apply the lessons accordingly.

Either way, though, regardless of what we call the emotion, it is not the problem. It's our response to that emotion that causes the worst mental ailment of all: second dart syndrome.

The Myth of Free Will

"Almost the entirety of what happens in your mental life is not under your conscious control."

—Dr. David Eagleman

Think of someone you hate or disagree with. What if you were born from that person's parents with those exact genes? And what if you had the same experiences of life? Would you be any different? Did he or she even have a choice to become that person? Did you have a choice to become who you are today?

Take a more routine example to illustrate the nature of free will: think about what it's like driving to work. Do you have to consciously process the route, or does the journey seem to go by on autopilot? This kind of automatic processing shows up everywhere because, like it or not, you can't control most of what happens in your brain. Almost all your thoughts and emotions are simply the result of conditioning and habit. As neuroscientist Dr. Jeffrey Schwartz says, "You really can't decide or determine what will initially grab your attention—your brain does." Our animal brain makes the decision without our conscious approval.

How does it decide? It chooses our reaction to events based on past patterns, implicit memory, genes, and pretty much everything that has brought you to where you are today. Dr. Benjamin Libet demonstrated our lack of free will by using EEG to prove activity could be detected in the brain's motor cortex close to three hundred milliseconds before a person consciously made the decision to move.

In another experiment, researchers found two separate areas of the brain held information on their participants' decision to hit a button a whopping seven to ten seconds before that decision was made consciously. Another group of researchers was able to predict their participants' movements seven hundred milliseconds before it occurred with up to 80 percent accuracy by tracking the activity in the brain.

Gerard Zaltman, a business school professor of Harvard, found that 95 percent of cognition happens in our subconscious brain. You might think you are consciously choosing to pick up your cup of coffee and have the freedom to do so, but your brain has made this decision before you even know it.

In his book, *The Happiness Hypothesis,* psychologist Jonathan Haidt uses the analogy of a rider and an elephant. Our human brain is the weaker, but smarter, rider sitting on top of the powerful, but less intelligent, elephant—the animal brain. We believe our human brain, the part we think of as our "self," is in charge, but in reality, our animal brain runs the show. If the elephant wants to go left and the rider wants to go right, who do you think is going to win that battle?

This is why so many people struggle with changing their behaviors, even though they know they want to. When someone struggling to lose weight sees a candy bar, it gets the less intelligent elephant all riled up. Buying that candy then becomes an automatic response to an external trigger, beyond the dieter's control.

The myth of free will isn't just a truth validated by science; this process is described in ancient spiritual teachings as well. According to Buddhist philosophy, when we experience any pain or suffering, it is the result of two darts. First darts are the ones beyond our control. They can be an external event or our automatic, internal response to them, based on the habitual patterns in our subconscious brain. Second darts are the manner in which we react and respond to the first ones. When you stub your toe against a door, the first dart is the pain in your toe. The second dart is when you say things like, "I am stupid," "This door is stupid," or "Why does God do this to me?"

On a trip up Bell Rock in Sedona, Arizona, my wife experienced firsthand the power of the two darts. Halfway up the rock, she consistently found herself too nervous to move. On certain exposed sections of the scramble, she clung to the wall with a force unbecoming of a five-foot-tall, eighty-five-pound woman. Within a few seconds, though, she would avert her eyes from the sharp drop to her right, look straight at me, gain a Zen-like focus, and keep moving forward, one step at a time. Eventually, she reached the highest point of the climb.

Getting back down proved to be much less of a challenge, and within the hour, we were in the car heading back home. While I was impressed with her first foray into such dizzying heights, she was angry with herself. In the comfort of the car, the first dart of fear dissipated, and the second dart struck.

As a climber, my brain processed the event differently because it had a greater number of reference points to pull from. She did not have the benefit of experience, so her brain was simply doing what it needed to keep her alive.

I wasn't any stronger than her. In fact, it took greater strength for her to climb the rock because she was afraid. I had no fear during the climb, so it required no courage on my part. Courage cannot be exercised unless fear stands in our way. As Nelson Mandela said, "Courage is not the absence of fear, but the triumph over it." Fear was the first dart, and was beyond her control. The second dart was a choice she had to make. She could choose to make the experience mean she was weak for experiencing fear, or she could choose it to mean she was brave for climbing higher despite it.

In life, there are always only two choices: one that empowers us to move toward our goals and one that disempowers us and brings us down to a life of mediocrity. The worst part is that one disempowering choice often triggers another one. First my wife felt afraid. Then she felt upset for being afraid, which led to more fear and worry about her own abilities, which led to a decreased sense of self-worth, and so on. "Second darts often trigger more second darts through associative neural networks," writes Dr. Rick Hanson. I call this second dart syndrome.

I once worked with a client whom I will call Bob. He suffered from severe anxiety every time he sat down at the computer to write. His palms got sweaty; his heart beat faster; his body trembled. As a result, he would retreat to his bed and turn on the television. It was a much easier way to silence his mind. Many therapists worked with him to get rid of the anxiety. When he came to me, I told him, "We are not going to get rid of the anxiety." That was not our goal, at least not initially.

The problem for him was not the anxiety; that was a pattern he had no control over anymore. His subconscious brain was the one responsible. The real problem was he thought of himself as weak for feeling anxiety just from sitting at a computer. He believed he was someone who would never succeed, which only fueled his love for the television, which then led to more feelings of worthlessness, which led to depression . . . You can see where this goes.

When I first started my business and a client said no to working with me, my thought pattern would go something like this: *Maybe I am not a good coach. What if no one hires me? Did I quit my job too soon? What will I do for money? How will I put food on the table?* I have intentionally beaten down this point because second dart syndrome is perhaps the worst mental ailment in the human condition. Letting go of our second darts will allow us to address all the other ones.

In all these examples, we were able to stop ourselves from sinking into second dart syndrome by first becoming aware of the fact that we had no control over it. Only then could we step outside of those patterns to choose new ones. Ironically, the act of surrendering control is what allows us to regain it. In the case of my wife, without accepting the limitations of her free will, she would have integrated the fear and the meaning she assigned to it into her identity. We all do this. We relate to ourselves by our emotions and our thoughts. They become a part of our sense of self.

When a young girl I know went into therapy, the therapist labeled her with the diagnosis of depression. She then believed that diagnosis was a part of her self-identity. The state of depression she experienced at times was no longer a passing emotion, separate from her sense of self; it now defined her. She would say things like, "I am depressed," or "I have depression," as opposed to something more accurate like, "My brain experiences a state of depression from time to time, but I am not my brain, and it is not me."

Bob did the same thing, until he began letting go of the second darts by accepting he had no control over his current state of anxiety. This is the most important first step in creating long-lasting change. It might be hard to accept your free will is limited, but only by accepting this truth will you then be able to take the necessary steps to regain your autonomy.

If you have a house that makes you happy, would you spend the time, money, and energy to go buy another one? Probably not. Similarly, if you believe yourself to be a fully autonomous being with free will who operates purely from a place of consciousness, why would you take any steps to realizing your autonomy, free will, and consciousness?

"Man is a machine, but a very peculiar machine," writes P.D. Ouspensky in *The Psychology of Man's Possible Evolution*. "He is a machine which, in right circumstances, and with right treatment, can know that he is a machine, and having fully realized this, he may find the ways to cease to be a machine." All it takes is the simple act of acceptance about our machine-like state to stop being a machine and start the process of mastery over the self. That should be easy. You get to accept the voice of the inner critic is not you. You get to stop making yourself guilty, labeling yourself as wrong or beating yourself up for everything that has occurred, externally and internally. You get to stop judging yourself. It's all not your fault.

While you have no control over this voice that often beats you down, with a little practice, you do have the power to condition even the subconscious parts of your brain. I developed the ability to climb Bell Rock without fear through conditioning. You too can train your mental muscles to process external stimuli in a manner that serves you.

Once you develop the ability to adapt your mind and self to the environment, whenever you are in unfamiliar territory and fear shows up, you will have the strength to keep putting one foot in front of the other. Before we build these fear-mastery muscles, let's take a look at how they are formed.

The Origins of Fear

"Just because there's that funny feeling in your belly doesn't mean that there's any threat. Our internal signals are pretty much bullshitting us all day long."
—Dr. Rick Hanson

Evolutionarily speaking, there are two causes of fear: learned and genetic. Certain fears are evolutionary constructs created to ensure our survival. For example, the Inuit people who live near the Arctic Circle have never seen a snake and, therefore, have not learned to fear them. Yet they get sweaty palms when they stare into the eyes of a slithery serpent. I would imagine there must have been a large threat of death by snakebite in the Serengeti, where it is said human beings first evolved into who we are today. Inevitably, evolution built in this fear to protect our species. However, not all of us are born with a fear of snakes. I have loved them since I was a child. Some of us are born with other fears.

In one study, psychologists placed babies on a table with a transparent piece of glass separating it from another one. The infants could have safely crossed the glass to the other table, but almost all of them chose not to. The clear glass triggered an innate fear of heights.

No matter what the fear, it has some evolutionary construct that made it valuable for survival. A fear of open spaces would make sense for our ancestors as it meant being exposed to predators. At the same time, a fear of tight spaces also made sense. That could mean no escape route.

Genetic fears are unique to each individual. There is no way to determine who will be born with what fear. Ultimately, though, as you might gather by now, it doesn't matter. Fear is just the first dart. If we choose to embrace it, it can be harnessed to make us stronger.

Then there are learned fears. Most Americans remember where they were on September 11, 2001. Images of planes crashing into the Twin Towers were plastered all over media outlets across the nation. People received the message loud and clear, or at least their brains did. They learned that planes equal death by terrorism, so they avoided them like the plague.

By fleeing from the airports, they killed themselves instead in car wrecks around the country. In the year after 9/11, German psychologist Gerd Gigerenzer found the number of vehicle fatalities skyrocketed. It stayed that way for a full year before the number returned to the norm from previous yearly averages. He even calculated that, as a direct result of people's decision to drive instead of fly, 1,595 Americans lost their lives. That is more than half the number of people killed in the terrorist attack on 9/11.

So what exactly happened here? The risk of dying in a car has always been higher than that of dying in a plane, before and after 9/11. Yet, Americans did not and are not fleeing from their vehicles because driving has not become a learned fear. Most Americans are blissfully unaware when a car wreck occurs in the country. They know nothing about it, so they have no fear about the risk it entails. Even if they did know the numbers associated with car wrecks, it would still not trigger a fear response. The risk of driving a car is commonplace; it has not triggered us on an emotional level. As a result, our brains deem driving a safe activity unworthy of fear.

On the contrary, almost all Americans knew about the events that took place on 9/11. Watching those planes crash into the World Trade Center and witnessing the horrors that followed as people leapt from burning buildings inevitably triggered our emotions. The strong emotional impact activated the amygdala, which embeds the fear response deeper into our memory.

This fear eventually dissipated once the act of flying became normalized again by a brain that no longer received any input indicating it was a risk worth heeding. Without the reinforcement of this learned fear, the memory faded and Americans returned to the airports.

Looking back, is it anyone's fault Americans responded to 9/11 in the manner they did? From the perspective of the animal brain, their fears were perfectly "rational," since the animal brain doesn't possess the ability to deem what is rational. You don't think of your dog as irrational if he chases a ball every time you throw it, do you? Similarly, their animal brain did exactly what one would expect from it, which is the definition of rationality. It generated what it thought was a reasonable and sensible response, based on the cognitive biases you will learn in the next chapter. We all do this.

"The optimal way of viewing the brain's ways of processing information may not be rational or irrational any more than the workings of the spleen or the liver might be viewed as rational or irrational. They are what they are and do what they do," writes Professor Michael McGuire in his book *Believing: The Neuroscience of Fantasies, Fears, and Convictions*. This goes against much of the conventional wisdom on fear that labels some fears as irrational or unreasonable. But the activation of the fear response does not possess the ability to be rational or reasonable, any more than an animal does. It occurs outside of our consciousness.

The act of judging our fears as unreasonable sets us up for long-term failure. It weakens our ability to face them by sending us down the spiral of second dart syndrome. If fear shows up, on some primal level, there is a valid neurological and psychological reason for it. Your animal brain does not know whether the object of the fear poses a legitimate threat to your life, so it responds based on your past experiences, beliefs, values, genetics, and your state of being.

I briefly worked with one woman whom we will call Amanda. She was scared to get married, despite knowing she loved her boyfriend and wanted to start a family. She thought it was "absurd" that she felt afraid. By digging deeper, it turned out her parents had a terrible marriage. Inevitably, her brain used that as a reference for what marriage looks like.

We shape our beliefs and perspectives from the information given by the world around us. In understanding this, she accepted the fear as normal and reasonable. However, patterns don't shift instantly. When the fear showed up again, she continued to deem it absurd and then began beating herself up for thinking of it as absurd.

We must not only use our human brain to interrupt second dart syndrome, but to understand the seed that implanted the fear as well. Without this level of awareness, we become victims to the world around us that is constantly teaching our brains what to fear.

Since we don't control what our amygdala learns to be afraid of, politicians, brands, "experts," and just about anyone can feed, manipulate, and exploit our fears for their own gain. For example, a 2008 study by the University of Bath in the United Kingdom found the strongest force of influence over consumers was exposing them to a terrifying version of their future self.

It is up to us to step outside our fears and look at them from a place of awareness to determine whether or not they are in service of our personal growth. Without this conscious assessment, our brains can even learn fear from fiction.

After watching the end-of-the-world disaster film *The Day after Tomorrow*, psychologist Anthony Leiserowitz found moviegoers were more concerned about global warming than before they saw it. People who had seen the movie even rated events, such as food shortages, nationwide flooding, and a new ice age, as more likely to occur than those who had not seen the movie. The movie is not only false, but absolutely ludicrous by any rational standards, yet it inspired fear in people.

The human brain and the animal brain often have very different reactions to events. As Jaimal Yogis, author of *The Fear Project* says, "What we feel afraid of is not always what we think we're afraid of." This is what makes fear so hard to face.

Let's say you are afraid of public speaking. Considering this is rated as the highest of all fears in a varied number of polls, this is a good example to use. You might consciously acknowledge to yourself there is no real reason to be afraid of standing up on a stage, speaking to a small group of people. You might even create a detailed list of all the scary things about that experience and rationally process why every single one of them is not worth being afraid of. Yet you cannot, for the life of you, figure out why you are still afraid.

This happened to me when I went skydiving. Despite having already completed three jumps, I was terrified of making the fourth. I consciously knew the odds of both my parachutes failing were extremely low. I wasn't scared of falling to my death. That didn't consciously occur to me as a potential risk at all, yet I somehow felt afraid. I remember driving to the drop zone with a part of me hoping it would rain so I could cancel the jump. Something about that plane door opening, the noise of the wind, and the blast of it on my face terrified me.

Just because your human brain sees no reason to fear doesn't mean your animal brain agrees. No fear is irrational or unreasonable. Fear is your friend when you choose it to be. You have no control over it anyway, so why fight it? Embrace it. I believed facing my fear of skydiving would strengthen my courage muscles, so I simply accepted the fear, regardless of why it showed up. I breathed it in, moved toward it, and leapt out of that plane.

Training Exercise

"You can be free from fear if you realize that fear is not the ogre."
—**Chogyam Trungpa**

The exercise below will help you become more self-aware about your fears. As I go through each action step, I will share examples from my life so you have an idea of how to apply the steps to your life. This will also help shatter the notion that someone writing a book about fear doesn't feel any fear. By revealing my own fears, I hope to help you create a new relationship with yours.

Step 1: Write down the events in your life causing you fear, stress, or anxiety. You can also start with a goal and write down the fears keeping you from taking action toward that goal. For example, my greatest fear at the time of this writing was the act of writing this book.

Step 2: Write out the thoughts, feelings, hidden fears, and terrifying possible consequences behind these stressors in your life. Dig deep here to find why you are afraid, how those fears are stopping you, and what is holding you back. Think of all the things that could go wrong if you fail to reach your desired goal and write them down.

A great exercise to help you find the subconscious meaning behind your fears is Toyota's 5 Whys technique. The method is simple: ask yourself the why behind your initial stressor at least five times. You could go beyond that number if you need to, but five whys generally gets to the underlying issue. This is what it looked like for me:

- Why does writing your book cause you fear and stress? *It takes a lot of time, focus, and effort, which doesn't always lead to a tangible result. (I would often spend all night writing and barely have a page to show for it.)*
- Why is that scary/stressful? *I have no idea if this effort will pay off.*
- Why does that scare you? *Committing to the hard work actually means I am committing to revealing my thoughts, beliefs, ideas, and all of myself to the world, leaving me exposed to far greater scrutiny.*
- Why is that scary? *People might judge me, hate my ideas, or leave bad reviews about the book. It could lead to criticism on a large scale.*
- Why does that scare you? *Maybe I will find I don't know enough or am not good enough to be "out there."*

You can see how my fears stemmed from something much deeper than just hard work. Almost always you will find this to be true.

To help you figure out what might be trapped inside of you, here are some common hidden fears that show up for many people:

- Fear of not being good enough: This can show up as not knowing enough, not being worthy enough, not deserving enough, or any variation of that idea.
- Fear of being alone
- Fear of leaving loved ones behind on your way to the top
- Fear of failure
- Fear of success
- Fear of the unknown: This is an obvious one, but it is worth exploring. Sometimes people have stories hidden in their subconscious that make them more prone to fearing the unknown than others.
- Fear of looking bad in front of others
- Fear of losing your freedom in some capacity

Fears can often start with some sort of "what if" scenario, such as, what if no one likes my book? So start with the words *what if*, and see what fears show up. The point of this step is to rationally process all elements of your fears: the why, what, and how.

Step 3: Write down the possibilities that lie on the other side of your fears. Clarity on the reward you will gain from taking action is an essential weapon in facing your fears. In my case, it was writing a *New York Times* bestselling book that will help millions of people face their fears and unleash their greatness onto the world.

Step 4: The simple act of bringing your subconscious fears into your awareness will allow you to alleviate their negative effect by choosing how you respond to them. You will learn more specific strategies on how to do that throughout the book, but for now, I want you to use your own creativity to brainstorm one to three solutions for the obstacles holding you back. Write down things you can do to prevent the negative outcomes you are afraid of. Create specific actions to address your fears. For example, I could study how to write a bestselling book.

Step 5: Preemptively prepare for what you will do should your fears become a reality. Write down actions for how you will respond to the possible worst case scenario. For example, if my book doesn't immediately become a bestseller, I could take a series of five specific marketing actions every single day until it does.

Jack Canfield did this for fourteen months to make *Chicken Soup for the Soul* a *New York Times* bestseller.

Step 6: Write down a list of Courage Transference Actions (CTAs). These are fear-inducing, confidence-building exercises that will help you transfer the effect of self-trust and confidence into other areas of your life. These must be actions you feel will help you grow psychologically stronger, not just actions that scare you for the sake of scaring you. For example, I still get creeped out by spiders, but I see no value in engaging that fear. On the other hand, I used to be—and still am a little—scared of heights. I could live my whole life without having to deal with that fear, but I choose to engage it by going skydiving, rock climbing, and bungee jumping. I can say with 100 percent certainty that taking on these fears has made me a stronger person in other areas of my life, so I continue to pursue such actions.

There is no right or wrong here. Perhaps conquering your fear of spiders will be useful for you. Use your growing self-awareness to decide what will be of value for you in building your courage muscles. Once you have a list you are satisfied with, take action and begin pursuing the items on that list.

Step 7: Write down past fears you have overcome and how you faced them. Take note of all the times you triumphed over any kind of struggle. What beliefs, strengths, and resources did you draw on to unleash your greatest self? For extra credit, write down all the achievements you are proud of in your life as well. The purpose of this list is to build references to show your brain you do have the strength and ability to face anything that comes in your way. More often than not, we do not acknowledge ourselves or celebrate our strengths, so this is the time to go big and brag about how amazing you are. You can even do this visually if you like and create a collage of images celebrating your past accomplishments.

Step 8: Write down stories about how you used fear, stress, or anxiety as positive forces to drive you into action. This will start to create a new belief about the nature of these emotions. When I did this exercise with a student I worked with, she realized the stress of having to complete a paper by the next morning helped her get it done faster than she normally would have. Looking back on that scenario helped her realize she could choose to make stress her ally any time she wanted.

The key takeaway from this chapter is to start practicing the belief that your fears are whatever you make them. You can use them to your benefit or let them beat you. As Lionel Richie said, "Greatness comes from fear. Fear can either shut us down and we go home, or we fight through it." The choice is yours.

Chapter 2

WHY WE DO THE
THINGS WE DO

"The first step is always self-awareness; without it, we're flying blind."
—Dr. Rick Hanson

I love annoying my wife. Among the techniques I use to annoy her, my favorite is putting my hands on her stomach—after I have held ice for a few minutes. This is generally followed by me scampering away, giggling like a child as she chases after me.

I am more needy and clingy than my wife, which makes me susceptible to simple jolts of positive reinforcement. I annoy her. She laughs. I do it again. It's similar to how we trained our puppy. He pees outside. We give him a treat. He does it again.

Just like our puppy, I get excited to see her every time she walks through the front door. She has now become acclimatized to two creatures swarming her, no matter how short her time away. One day after she finished two hours of yoga and meditation, we both ran over to greet her. On this particular occasion, just as I was about to give her a big hug, she warned me, "Don't touch my back; it hurts." Guess what I did?

I should probably clarify that I love my wife, and she loves me. We have a happy marriage after five years. Our pranks are just playful ways to make each other laugh. I would never intentionally hurt her. Yet, before I knew what I was doing, my hands went straight to her back.

Everything we do at every moment teaches our brain how to react to the next one. Almost all of this training occurs subconsciously. My brain has learned annoying my wife equals pleasure, so when she asked me not to do something, it responded automatically by doing that very thing. This is one small example of how our animal brain runs the show.

After delving into subjects like suicide and war, I thought these pranks might lighten the mood a little bit, but my playful antics also highlight the kind of automatic behavior that occurs throughout our lives. Charles Duhigg, author of *The Power of Habit*, says that 40 to 45 percent of what we do every day is just a habit.

Even the actions we seem to be choosing consciously are often a result of a subconscious decision driven by an emotional need within the animal brain. "The thoughts and actions that System 2 (your human brain) believes it has chosen are often guided by the figure at the center of the story, System 1 (your animal brain)," states Nobel Prize—winning psychologist Dr. Daniel Kahneman.

Remember the myth of free will?

People all over the world engage in activities they consciously recognize as not beneficial to their growth. People smoke cigarettes and drink alcohol, knowing it damages their health. The average American spends five hours a day watching television. Why would we spend so much time involved in activities we know will not improve our lives?

At its core, there are only two reasons for why we do anything: to avoid pain or gain pleasure. These are the two driving forces of all human behavior. The things we do automatically have been conditioned to mean pleasure inside our brain.

To become a peak performer, our goal is to bring awareness and focus to how our brain learns the meaning of pain and pleasure. This doesn't mean we want to stop to think about every action we take. That would be a waste of precious mental energy and also highly unproductive, even dysfunctional. What we do want is to leverage our human brain to consciously choose what behaviors become automatic and which ones equate to pleasure. To do this, we must first know how our animal brain makes these choices to begin with.

The following is not an exhaustive list of cognitive biases, the term used to describe the various functions within our animal brain. I have chosen a few I believe to be relevant for this subject matter. If you want to dig deep into the inner workings of the animal brain, you will find a detailed list of twenty-five cognitive biases at www.fearvana.com/resources.

The Law of Love and Hate

How would you feel if I told you that someone would be installing a nuclear power plant in your city later this week? I would guess you are not a big fan of the idea. But what if I told you nuclear power produces almost no air pollution, is a long-term sustainable source of energy, and reduces carbon emissions? In fact, according to Professor Bernard Cohen, even if we were to take into account the probability of every nuclear accident happening, it would "eventually expose the average American to about 0.2 percent of his exposure from natural radiation."

This minimal level of radiation reduces our life expectancy by less than one hour, whereas coal, gas, oil, and electric generation technologies reduce our life expectancy by anywhere from three to forty days. In fact, most experts on the subject assert nuclear power is not nearly as dangerous as we tend to think. Does any of this make you feel more comfortable about a nuclear power plant in your city? Probably not. Why? Because we operate from feelings.

All life is first filtered through the emotional animal brain before it reaches the human brain. Our animal brain remembers the Chernobyl disaster. That incident became a reference for millions of people all over the world, telling their brains to dislike nuclear power.

When we hate something, no matter how many facts may disagree with our emotions about that thing, our final opinion will be based on feeling, not reason. Psychologist Paul Slovic calls this the affect heuristic. I like to call it the law of love and hate. It states that our gut feelings of love and hate determine our decisions and our views of the world.

Slovic demonstrated this in a study that showed how participants changed their beliefs about the risks of various technologies, such as chemical plants or food preservatives, once exposed to its benefits. The participants only received emotionally compelling evidence about the value of these technologies, but received no evidence whatsoever about their risks. Yet, once their animal brain fell in love with the technology, their beliefs about its risk changed as well.

This is why I created the concept of Fearvana. Fear has a negative connotation. The law of love and hate tells your animal brain that you hate fear. This emotion is deeply ingrained, so you will continue to resist fear, despite the rational scientific evidence proving resistance to be futile. With Fearvana, the law of love and hate has no context or reference point. We can now start from scratch and fall in love with Fearvana.

The Negativity Bias

Picture this: Our ancient ancestors are going about their day in the Serengeti, searching for food to survive. One day, they miss out on a kill. No big deal. There will be plenty of animals to hunt the next day. They go back to their homes, admiring the beautiful sunset beyond the plains. But if they fail to notice the saber-tooth tiger waiting for them around the corner, they will have no need for food—forever!

What does this mean for you today? Psychology professor John Cacioppo conducted a study in which he showed a group of people three kinds of pictures while recording the electric activity in their brain. He showed them pictures known to arouse positive emotions, such as a Ferrari or a pizza; pictures that elicit negative emotions, such as a mutilated face or a dead cat; and pictures that arouse neutral emotions, such as a plate or a hair dryer. He found that the electrical activity in the brain increased when it was exposed to negative stimuli.

Dr. Roy Baumeister, in another study, discovered participants expressed greater upset about losing fifty dollars than joy over gaining the same amount.

We are more heavily influenced by bad news than good news. Turn on any news station as evidence for our negativity bias.

Researchers have found this negativity bias exists even in three year olds. It is not something we create in our minds as adults; it is an instinctual condition. Because of this evolutionary instinct for survival, we pay more attention to the threats around us, as opposed to the beauty. This is why most people focus on what they don't have, not on what they do have, and we act accordingly as well. "Our brains have become like Velcro for negative experiences and Teflon for positive ones," says neuropsychologist Rick Hanson.

The Top-of-Mind Rule

In studies conducted on public perceptions of risk, psychology professor Paul Slovic found that people's estimations were far from the facts. People felt tornadoes kill more people than asthma. The reality is that asthma takes twenty times the number of lives as tornadoes. Eighty percent of people deemed accidental death more likely than death by stroke, despite the reality that strokes cause nearly twice as many deaths as all the different kinds of accidental deaths combined.

Perception of risk is drastically altered by the media; the unrelenting coverage puts certain events at the top of our minds. This exposure shapes our beliefs about the world, and how our animal brain responds to it.

The easier it is to recall something from memory, the more likely our brain deems it important or commonplace. This psychological condition is known as the availability heuristic or, as I like to call it, the top-of-mind rule. Whatever is on the top of your mind is more likely to have an impact on your perception of reality. This makes sense evolutionarily speaking. If two of my buddies told me to be careful when I went to the lake for water because a monster crocodile happened to be lurking there, I would want that to be at the top of my mind, right?

As Kahneman states, "The world in our heads is not a precise replica of reality; our expectations about the frequency of events are distorted by the prevalence and emotional intensity of the messages to which we are exposed." To make the top-of-mind rule work for us instead of against us, we need to take control of the messages we are exposed to and consciously decide what we want to be at the top of our mind.

Anchoring

In a study conducted by Kahneman and Tversky, they instructed participants to spin a wheel with numbers ranging from zero to one hundred, and write down the number it landed on. (They didn't realize, however, that Kahneman and Tversky had rigged the wheel to stop only on number ten or number sixty-five.) Afterward, they asked two questions: Is the percentage of African nations among UN members larger or smaller than the number you just wrote? What is your best guess of the percentage of African nations in the UN?

The average guess of the participants who landed on ten was 25 percent, while the average guess of those who landed on sixty-five was 45 percent. Despite the fact that a spin of a wheel provides nothing of value in terms of guessing the percentage of African nations in the UN, those who spun a lower number estimated a lower percentage. The number acted like an anchor from which to make their decision.

In another study, a group of psychologists placed a sign that read "limit twelve per customer" over cans of tomato soup. They found that when the sign was in place, most customers bought anywhere from four to ten cans. Not a single customer bought one or two cans. When the sign was taken down, nearly half the shoppers bought only one or two cans. Just seeing the number twelve placed an anchor in the shoppers' minds that shaped their behavior. Without awareness and acceptance, the external world chooses for us, without us having a say in the matter.

This doesn't just occur with numbers either; anchors can take any form. To this day, the song "When You Say Nothing at All" by Alison Krauss reminds me of my first serious girlfriend. During our relationship, the song formed an anchor in my memory that remains active. Anchors can show up as songs, movies, places, people, events—almost anything. They form deep roots, especially when they are combined with emotionally charged experiences.

Anchors can victimize us to information acquired beyond our control, like a supermarket sign, but when used consciously, they have the power to generate new psychological and physical actions at will. For example, when I overcome pain on a long run, I stay present to that inner power and anchor it into me so I can call forth that strength when I need it again later.

Confirmation Bias

Have you ever gotten into an argument with someone, and no matter what you say, no matter how much evidence you have to back up your point, you just can't convince him you are right? You might have had this experience with your spouse—maybe just once or twice. Once a belief is planted in our minds, we commit to it entirely. We filter out external information, based on whether or not it is in line with our beliefs, and then act according to those beliefs.

In an experiment conducted at Stanford University, a team of researchers brought together two groups of people passionate about capital punishment. Half the group was in favor of it while the other half was against it. Both groups were presented with fictional statistics from studies comparing states with and without the death penalty. Then they were presented with more detailed information about how the studies had been conducted. Each group sought out evidence to validate their viewpoint, while claiming any evidence that pointed to the contrary was insufficient or irrelevant. They screened the information to ensure it was consistent with their preexisting beliefs about capital punishment.

The effects of confirmation bias can be seen clearly during election season. Whether people label themselves as Democrats or Republicans, they will only find evidence to justify their chosen party is correct in almost every way and the other is wrong. They will even interpret any story in a manner that supports the decisions of their candidates. We all do this for every one of our beliefs.

Research with MRI scanners demonstrated the emotional brain lights up when we are presented with information contradicting our beliefs. We stay safe and comfortable in our beliefs, so we hold on to them and filter out any information that contradicts our views. Our animal brain doesn't like things that threaten the security of our internal or external world. Confirmation bias is why we often engage in an activity to satisfy an emotional desire, then find a reason to defend the action.

Referential Thinking

Your brain processes the world around you based on references it receives from all external stimuli. Everything you have ever experienced in your life shapes the references you form. These then create the patterns that shape your behavior.

A marathon might feel like a walk in the park to one person and the toughest thing in the world to another. Through training, one person can develop a reference point the other does not have. When I felt no fear while climbing in Sedona, that was because of the references my animal brain had created from past climbing experiences. My wife's brain did not have such references.

I witnessed a powerful example of the impact referential thinking has on our ability to process fear while traveling through Israel. During one of my visits, a bomb was found in a mall near my parents' home. The next day, everyone was back in the mall and on the streets. The same phenomenon occurred when war broke out in neighboring Lebanon. In a country that is in a constant state of threat from their neighbors and where everyone serves in the military, the quantity and quality of their references allow them to respond with calm to external events, like war and potential explosions. The references we form help us develop a sense of how our world works so we can function effectively in it without constantly being on a state of high alert.

Referential thinking affects your opinion on everything around you as well. The reason I love my KitchenAid blender is because I view it in reference to all the other blenders out there. (I was not paid to make this endorsement; I just happen to love my blender.) Psychology professor Dr. Dan Ariely says, "We are always looking at the things around us in relation to others. We can't help it. This holds true not only for physical things but for experiences such as vacations and educational options, and for ephemeral things as well: emotions, attitudes, and points of view."

Law of Perceived Commonality

Despite a conscious admission of not being racist, studies have shown many white people register a stronger fear response in their amygdala when they see a black face than when they see a white face. Why? Through extensive media exposure, our animal brain believes a black person is more likely to be a criminal than a white person. If it is perceived to be more common that a criminal would be a black man, then it is perceived to be more common that a black man is a criminal. Remember, your human brain and animal brain don't always agree on what is scary.

According to the law of perceived commonality, we make judgments on something based on our belief of what is commonly known about that thing. Our animal brain created the law of perceived commonality as a means of survival. If lions are commonly seen in tall, dry grass, then tall, dry grass must commonly have a lion hidden it. One would be wise to stay away.

What Can You Do about All of This?

"Awareness of the power of the unconscious to guide and change our thinking and behavior is the first step to change. If we deny how much of our thought and behavior is unconscious, we'll have less chance of making changes stick."
—Dr. Jeremy Dean

Here is an example of how these cognitive biases work together to implant fear into our head. Before leaving for my run across Barbados as a part of my goal to run across every country in the world, I read an article about a tourist who was robbed at gunpoint. The negativity bias ensured that out of everything I knew about the country, this was the only fact etched in my brain. Once I began my run in an unfamiliar environment, this was then at the top of my mind. A few miles in, I got lost. Guess what fearful thought ran through my head?

The law of love and hate created a belief that made my brain respond to something as simple as asking for directions with anxiety. Having had enough practice in fear, I interrupted the descent into second dart syndrome and consciously took the risk to ask someone how to get back to my chosen route.

The man I asked proved to be extremely helpful and kind. My brain just found a new reference to eliminate the fear memory imprinted by the article. When I stopped a second time to confirm I was on the right path, the cop was even friendlier than the first person. That sealed a whole new belief about Barbados and its people. The law of perceived commonality now told my brain that since two locals were friendly and kind, all locals must be the same way. I then made the conscious choice to anchor this belief about the people of Barbados, so when I met a stranger later on in the run, I gladly stopped and asked him to take a picture of me running.

Evolution created these operational procedures to simplify a complicated world around us and keep us alive. We need them to function effectively. They allow us to rapidly respond to events without burdening our slower and lazier human brain. So our goal is not to eliminate the effects of these biases. We couldn't, even if we wanted to. Our animal brain creates patterns without our awareness. It is our human brain that gives meaning to those patterns to form new ones.

By utilizing your human brain to make conscious decisions, you can literally change the physical structure and functions of your animal brain. Your cognitive biases won't disappear, but you can change the way they operate for you.

The first step in the journey to mastery over the self is awareness. Now that you are aware of many of your biases, you have taken one step further away from a machine-like state of existence and one step closer to seizing control over your consciousness.

Training Exercise

The problem with many of our cognitive biases is they do not always serve us in a modern world our brain is not designed for. In this exercise, we will explore how they are affecting you on a day-to-day basis. I highly recommend you download the full list of twenty-five cognitive biases at www.fearvana.com/resources and do this for all twenty-five of them. However, if you choose not to, at least do this for the seven cognitive biases outlined in this chapter.

You just read about how cognitive biases affected my life. If it would help, use my experience as an example to think about or write down at least one event in your life where each of these biases has affected you. The goal is to become more mindful of their influence over you so that, in time, you can make them work for you instead of against you. Here are the seven cognitive biases again for your convenience:

- The law of love and hate
- The negativity bias
- The top-of-mind rule
- Anchoring
- Confirmation bias
- Referential thinking
- Law of perceived commonality

After thinking about or writing down one event, practice becoming aware of how theses biases impact you in every area of your life. Consciously notice their presence, and see whether their automatic responses are programmed in alignment with the person you want to be. As you do this, also notice what events, people, and environmental cues trigger a negative response and which generate a positive response. This will help you prepare for such situations and use your consciousness to reprogram your cognitive biases. For extra credit, use the 5 Whys technique from the last chapter to figure out why certain situations bother you when others do not.

Depending on where you are in your life and what challenges you are facing, this level of awareness might be more important or less important for you. At one point in my life, I needed to become extremely self-aware to figure out why I wanted to drink alcohol, but I did not need this level of awareness to exercise on a regular basis. I had become accustomed to facing my fears and choosing to suffer physically through exercise, so I did not want to waste cognitive energy. Instead, through exercise, I allowed myself to freely enter into Fearvana.

Since you are here to learn how to turn your internal obstacles into your allies, this exercise will be of some value to you, so I do recommend you practice it to whatever degree and in whichever area you see fit. Activating your consciousness throughout the day does require a lot of effort. As we move through our work together in this book, we will no longer need to expend this energy. Eventually, all your cognitive biases can and will automatically work in service of the greatest version of yourself. Imagine the power of being able to control what your mind does, even when you are not thinking about it. That is the power of a superhero.

Chapter 3

OUR ONLY REAL FREEDOM

"Our obligation is to give meaning to life and in doing so overcome the passive, indifferent life."

—Elie Wiesel

I didn't even have to look up. I felt him. I felt like I was going to vomit. As he approached the hostess stand, it was like nothing I have ever felt before. My whole body responded. I'm shaking. I'm hyperventilating. I immediately just knew inside, somehow knew that it was him. I had to run to the back of the restaurant, and I had a panic attack. I threw up and threw up and threw up."

As a six-year-old child, Alice lived down the street from a close friend, Mindy. Carefree Alice often spent hours over at her friend's house jump roping and playing games outside in the yard. Mindy's household included her mom, dad, little brother, and a friend of the father who seemed to be a regular guest. Every once in a while, Alice slept over at Mindy's house so they could play together for as long as possible. When it came time for bed, Alice climbed into the bottom bunk while Mindy clambered up the ladder to the top. But in that house, lights out did not mean lights out.

At some point in the night, the door made a faint, creaking sound. A "big, fat, gross man with a big, giant beer belly" pushed it open and crept into the room. Shutting the door behind him, he inched his way toward the bottom bunk. "He put his forearm across my mouth to keep me from making any sounds. Then he would sexually assault me."

For almost two years, Alice was brutally raped by that man on multiple occasions. Recalling her experience, Alice said, "I knew it was bad. I knew it hurt. I knew I hated it. But I didn't really have a name for it yet." How could she? She was six years old.

"Being so young when that happens, you don't have a reference for what's happening. It's not like I had sex before or even understood what it might be. There's no concept of anything entering me. He was hurting me for sure, but I don't know if I understood it as sexual," Alice said. The suffering she endured in that room carved out just one part of the inferno that raged all around her.

Throughout her childhood, Alice's father spent his time in and out of jail. Her mother was diagnosed with bipolar disorder and remained untreated. Her volatile condition left her unequipped to care for a child in every way imaginable. She was on drugs and "was emotionally very up and down," which made her "really tough to be around." Many years later, when Alice confronted her mother about the sexual abuse, she discovered that, on some level, her mother knew but pretended it didn't exist. She couldn't accept the reality of it, despite witnessing firsthand the harrowing consequences of the incident for young Alice.

"I would wake up screaming, drenched in sweat. I had these recurring dreams that I was kidnapped and made to watch my friend who was also kidnapped die. It was a horrible, horrible dream. And I had a variation of it where I was made to dress up in weird clothes. It was really scary." To her mother, these were just bad dreams, unrelated to her child's waking life, so she ignored them.

As her mother became harder and harder to live with, things got worse for Alice. When Alice turned fifteen, her mother kicked her out of the house. Forced out into the streets, Alice needed money to survive. Alone and desperate, she had to get a job. She found one as a hostess in a restaurant, where she was forced to confront the same man who raped her as a child.

Her nightmare didn't end there. Without the security of a loving home to keep her safe, Alice was once again raped. That time, she was gang raped by five teenage boys. That time she was more than old enough to understand what was happening.

Finding Joy in War

"Everything can be taken from a man but one thing: the last of human freedoms—to choose one's attitude in any given set of circumstances; to choose one's own way."

—**Victor Frankl**

Although horrifying and shocking, the most incredible part of Alice's journey was not the degree to which she suffered, but the manner in which she responded to it. "I wouldn't change anything about my past," Alice told me. "I think the only reason I can say that is because I don't see it as having ruined or taken anything away from my future. I really don't." More often than not, we can't control the obstacles life will place in our way. Alice certainly couldn't do anything to prevent what happened to her. Yet, she chose not to let her circumstances define her.

Today Alice is happily married, has a child on the way, and has a very successful professional life. The strength to create that life—despite all she suffered—came from "the ability to always look for another version of the story that actually allows me to move forward." She told me, "This is not the version of the story I am willing to live . . . I will be with this, but I will not be this. This cannot be it for me." Alice is one of the finest examples I have ever encountered of our collective ability to separate ourselves from suffering to create our own empowering reality. It took me a lot longer to learn that skill.

For the first two months of my deployment in Iraq, I hated my life. I spent my time alone, consumed with anger, rage, and frustration over the leaders, the missions, the misery, the lack of freedom—everything! I wrote this in my journal a month and a half into the war: "I find myself getting more and more agitated with the shit I am doing here, and consequently, I spend more time with myself retreating into my own world . . . I fear I am going to be angry and impatient when I return home . . . I find myself less motivated to wake up in the morning, even to work out or train, because I don't want to wake up to the shittiness of this reality."

Eventually, I learned how to change the meaning of the war by delving into the minds of philosophers and psychologists like Marcus Aurelius and Eric Fromm. I wrote the following words in my journal three months into my deployment: "Two people can live in the same place and at the same time, and one can be happy while the other is sad . . . My world is how I choose to adjust to the world around me. The place would be miserable only if I made it so."

During the war, I first learned how to isolate myself from my emotions to become aware of them from a higher place. This allowed me to find freedom, even in a place where I had none. "In the last day or two I have been indifferent to what we do, and, in turn, I am angry and frustrated less and laughing and smiling more. I have not been thinking about what we are doing and why. I just get up and do it, and I find that I am better off because of that. It is a situation beyond my control, so I should just accept it, adapt, and make the best of it. My mind has become free," I wrote.

Unlike life back home, where I had more freedom over what I did with my day, I had no control over my life in Iraq. As a Marine, I followed orders. All that mattered was the mission and the men beside me. If I was told to do something, I did it. That demanded a higher level of separation—to the point of indifference—in order not to let that environment affect me. I don't recommend such indifference in the "normal" world where you have more control over your life and your decisions, but at the time it served me well. Eventually, this skill allowed me find joy in everything the war threw at me.

"It's New Year's Eve 2007," I wrote in my journal later into my deployment. "I am sleeping on a floor in the basement of an abandoned train station in the middle of the Iraqi desert in winter, and I am strangely happy . . . I really am going to miss these good times when I go back home." By seizing control of my consciousness, I managed to change my automatic response to the war. As Dr. Hanson writes, "Over time through training and shaping your mind and brain, you can even change what arises, increasing what's positive and decreasing what's negative." Until we develop that internal mastery over our mind, however, any event could cause a great deal of fear, stress, or anxiety for some, while barely fazing others.

I believe the circumstances that brought me to the moment where I thought about committing suicide are trivial compared to the suffering of people like Alice and Dale. Yet I let my suffering devour me, and I almost lost the battle to it. On the other hand, Alice never let it consume her for more than a few days. This demonstrates the true nature of adversity.

Although hardship may be universal, how it is experienced is entirely relative to the individual facing it—regardless of the context that creates it. I have seen high school students in upper class neighborhoods suffering more in preparation for the SAT than a child with no money and food living on the streets of India. A celebrity with fame and fortune could easily experience more suffering than a woman who lost her child and is paralyzed from the neck down. (You will read about her later.)

What all of this means for you is that it does not matter what you might be struggling with or how it might compare to other people's suffering. Don't waste your time and energy in the destructive downward spiral of second dart syndrome. For many people, this is hard to grasp. Someone once responded to an article about Fearvana by saying, "When responsibilities, finances, and a lack of access to opportunity comes in, getting over fear and crossing over is a lot more difficult."

People often feel like the external reality determines our level of suffering, but as you have seen, none of those outside forces matter. What we do inside our minds, the conversation we have with ourselves, that is what shapes our reality. Our mindset determines how much we suffer. You have the power and the freedom to choose how you interpret your world. That interpretation will control the quality of your life. If you don't exercise this freedom, your brain will create its own interpretations without your awareness.

Psychology professor Dr. Michael Gazzaniga discovered this during his research with split-brain patients whose left brains were disconnected from their right brains. Without the corpus callosum, the nerves that connect the two hemispheres, it was impossible for the two brains to communicate with each other. Yet, somehow, the left brain found a reason to justify the actions of the right brain.

Within the left hemisphere of the human brain lies what Gazzaniga calls the interpreter module. It "continually explains the world using the inputs it has from the current cognitive state and cues from the surroundings." In one of his experiments, Gazzaniga flashed the word *red* to a patient's left hemisphere and *banana* to the right hemisphere. He then placed a set of colored pens in front of the patient and asked him to draw something with his left hand.

The patient's left brain controlled the decision to choose a color, so he picked up a red pen. The right brain controlled the decision to draw a banana with his left hand. When asked why he drew a banana, he replied, "It is the easiest to draw with this hand because this hand can pull down easier." Yet, the interpreter module in the patient's left brain, which could not have known why the left hand drew a banana, nonetheless compelled him to draw a banana. The correct answer to why he drew a banana should have been, "I don't know." Without any way for the two hemispheres to communicate, he could not have known why his right brain compelled his left hand to draw that banana.

"The interpretive mechanism of the left hemisphere is always hard at work, seeking the meaning of events. It is constantly looking for order and reason, even when there is none," said Gazzaniga. Your brain is finding meaning whether you like it or not, so you might as well take charge and choose a meaning that serves you. If you don't, you may not always like the one your brain finds on its own.

The only thing that matters for you is whether or not the meaning you consciously choose in response to your stressors drives you into action. You can use the examples above as references to find the value and growth in your own struggle, but I know this is easier said than done. Otherwise, fear wouldn't be such a formidable barrier, and everyone would know how to boldly face any challenge that comes their way.

Let's look at why that doesn't happen. By learning how the human brain functions in conjunction with the animal brain, you will then have the power to harness both of them to let go of those second darts.

The Curse and Gift of Being Human

"Life has no meaning. Each of us has meaning and we bring it to life. It is a waste to be asking the question when you are the answer."

—Joseph Campbell

Our human brain separates us from the animal by granting us self-awareness and consciousness. It gives you the means to read this book and also say to yourself, "I am reading this book." The human brain also allows you to process events and respond in a more deliberate manner than you would if the animal brain controlled the decision. It is slower than the animal brain precisely because it employs reason and rationality.

From language to the ancient pyramids to the newest iPad, every one of our creations since the beginning of time has been a product of the human brain. It took reason and logic to solve all the problems that threatened those creations. Ultimately the human brain shapes how we define ourselves and the world around us.

The human brain grants us the superhuman ability to mold our reality by assigning meaning to the external world and our automatic reactions to it, including all the cognitive biases that occur in the animal brain. However, we are not endowed with this gift. It must be earned through training and the constant pursuit of mastery over the self. Without such rigorous practice, we will fall victim to the side effects of self-awareness: criticism, doubt, negativity, and all the internal dialogue that holds us back. Those are the second darts aimed at our self-worth.

You don't see animals judging themselves for their behavior, do you? It is the human brain that prevents us from simply being by questioning everything we do. The animal brain is a doer and actor. The human brain is a resister and doubter. It creates excuses and justifies inaction. The paralysis of perfectionism so many of us experience in our quest for greatness is a direct result of this burden of rationality.

Inevitably then, turning off the human brain is something we all crave in one form or another. It can be through drugs, alcohol, television, adventure sports, surfing the Internet—anything to run away from having to be with ourselves. We seek this silence not only to shut off our inner critic, but also because the natural state of our mind is chaos.

In the moment it took to read that last line, your human brain was updated five to eight times. You have one hundred billion nerve cells and one hundred to five hundred trillion neural pathways that transport information within your brain. The number of possible combinations that can be formed is ten to the millionth power, far more than what we could experience in one lifetime. It is believed that this number is greater than the number of known particles in the universe.

It is a tumultuous world inside your head. Of course, you are going to want to turn off that chaos. To see what I mean, try this exercise. Set a timer for two minutes. Sit still with your eyes closed and think about nothing. Empty your mind of all thoughts for just two minutes. It's near impossible, isn't it?

Even if you did manage to achieve stillness of the mind, it probably did not last more than a few seconds, if that. Thoughts run wild inside our head. Chaos is normal. This makes it hard to focus and take action to accomplish the results we want. The overwhelming number of choices we have today only aggravates this natural state of our mind.

To make things worse, our chaotic mind works in conjunction with the negativity bias and the top-of-mind rule, so it naturally veers to whatever presents the greatest perceived danger, threat, or problem in the present moment. You are probably well aware that when the mind wanders, it drifts to the direction of all your fears and doubts. Collectively, all these features of the brain trigger the disease of second dart syndrome—unless we channel its power to interrupt that destructive pattern.

The Peak Performer's Secret

"Between stimulus and response there is a space. In that space is our power to choose our response. In our response lies our growth and our freedom."
—Victor Frankl

To navigate the challenges we face in life, we must learn when to let the animal brain take control and when to activate the human brain. There are times when we need to stop thinking and leap into action before we feel ready. There are also times when we must pause to consciously determine whether the chosen action is the right one to take.

It's probably clear by now that taking on this war against our limitations is no easy feat. Not only do we have no control over one half of the weapon, the other half is a chaotic mess that questions everything we want to do. On the plus side, however, one part is fast and efficient while the other is highly intelligent and creative. To harness the benefit of speed, with the power of reason, we need to make these two brains work together. This not only makes us better at what we do, it can also mean making more money.

After years of research, Travis Bradberry and Jean Graves found that 90 percent of high performers are high in emotional intelligence (EQ). They even found that people with high EQ make an average of $29,000 more in a year than those with low EQ.

Emotional intelligence is knowing when to step outside of your emotions to choose a rational response and when to let yourself be consumed by them. It is the ability to effectively manage the flow of communication between your animal brain and human brain. Think of it like information traveling across as a highway that connects the two brains. "The more you think about what you are feeling—and do something productive with that feeling—the more developed this pathway becomes," wrote Bradberry and Greaves.

In a 1993 study, researchers found that the most commonly used strategy for national champion figure skaters to handle the rigors of training was "rational thinking and self-talk." The figure skaters learned to step outside their feelings, examine them, and talk themselves through the challenges rationally. No matter what the emotion, it can be useful if we follow the same strategy and process it effectively using our human brain.

Before I went to Iraq, I lost a close friend and fellow Marine to the war. Simon and I both served in the same unit and were similar in our abilities as Marines. On the physical fitness test and the rifle range, our scores were almost identical, but I always beat him by a few points. We became close and wanted to go to war together. We volunteered for Iraq every chance we could, but it never happened.

One summer, I left to visit my family in India, and he found a unit to deploy with. Being an outstanding Marine, he was quickly promoted to corporal and put in charge of a vehicle team. In January 2007, an improvised explosive device struck Simon's Humvee, killing him instantly. To this day, I believe that if I went to war with him, I would have gotten that promotion. I would have been in that seat, and he would have come home to his fiancé.

For a long time, everyone told me all the reasons in the world why I should not feel guilty. Rationally, it made sense to me. Even if we had deployed together, a million things could have gone differently, and he still might have been killed. But emotionally, none of that mattered. Even as I write this, I feel the tears welling up in my eyes.

No matter what anyone ever said, it never made a difference until I taught myself everything I am sharing with you in this book. The guilt still lives within me, but I have found value in it. I now use it to do something meaningful. I tell myself since I am still alive, I cannot not waste this. Let me do something worthwhile with my life. I keep his picture on my wall to remind me I need to earn this life that should not have been mine. Guilt has become my fuel to live life to the fullest and, more importantly to use it in service of a greater good.

Similarly, research has found that other emotions traditionally thought of as negative, such as sadness and anger, also have value if we learn how to use them. When studying why people watch sad movies, associate professor of communication at Ohio State University, Silvia Knobloch-Westerwick, found that "positive emotions are generally a signal that everything is fine, you don't have to worry, you don't have to think about issues in your life. But negative emotions, like sadness, make you think more critically about your situation. So seeing a tragic movie about star-crossed lovers may make you sad, but that will cause you to think more about your own close relationships and appreciate them more."

During his research on the effects of anger, Craig Kennedy, a professor of special education and pediatrics, discovered "an individual will intentionally seek out an aggressive encounter solely because they experience a rewarding sensation from it. This shows for the first time that aggression, on its own, is motivating, and that the well-known positive reinforcer dopamine plays a critical role."

Contrary to popular belief, there are no good or bad emotions. All emotions can drive us forward or hold us back. Our job is to find the value in any emotion and leverage its energy. To take it one step further, nothing is inherently bad or good, whether it be an emotion, an experience, an event, a person, or the environment around us. The meaning we consciously assign to everything is what matters.

"You have these feelings, you have these emotions, what are you going to make them mean?" Alice told me during our interview. "I had refused to make it mean that because this happened when I was six or seven or eight that I can't have a really happy story and intimate marriage. I'm willing to relate to it as something I have an option for."

Alice learned how to use her human brain to consciously choose a new meaning for events that easily could have left her struggling with post-traumatic stress disorder for the rest of her life. However, to be able to relate to her story and her emotions that way, Alice needed to first become aware of them. "No attempt to stuff it works," she said. "I tried to just stuff it and pretend that everything was fine, and that did not work. I would not advise that."

Hiding or running away from our selves resigns our power to the realm of the subconscious, where we have no jurisdiction. Bringing the forces that drive our behavior into our awareness allows us to choose how we respond to them. "Self-awareness is a foundational skill; when you have it, self-awareness makes the other emotional intelligence skills much easier to use," Bradberry and Greaves wrote. They found that 83 percent of people skilled in self-awareness are peak performers in other areas of their life as well. This is why everything we have discussed up to this point is about building your self-awareness muscles. They give you the strength to navigate your way through fear into Fearvana.

Training Exercises
Here are four different strategies to help you improve the communication between your animal brain and human brain.

Practice Presence

Start paying more attention to your thoughts and emotions by asking yourself questions: What thoughts are running through my head right now? What am I feeling? Practice doing this by setting an alarm to ring every hour, or practice it every time you walk to work or whenever you are waiting in line. The goal here is to become better at noticing the automatic thoughts and emotional patterns that show up as a reaction to all life events, especially the challenging ones.

Keep a journal to track how you respond to such events. Log your emotions, your thoughts, the triggers that activate you into action and the ones that paralyze you, along with anything else you can use to get better at responding to the challenges life throws at you.

Being fully present to your self will help you control the flow of communication between your two brains. The more you become aware of your thoughts and emotions, the more power you have to step outside of them and choose something beyond them.

Let Go of Judging Your Emotions

We label some emotions as "bad" and others as "good" to simplify our world, but this does not help us. Judging our emotions leads to second dart syndrome. Guilt is often considered one of the "worst" emotions, yet I learned how to harness it to work for me by letting go of the judgment associated with it.

During my struggles with alcohol, as soon as I started the sober part of the cycle, I would beat myself up and call myself the worst names imaginable. It was a textbook case of second dart syndrome. This is horrific if you think about it. Most of us would never tell a struggling addict what a worthless human being he is. We would support him and love him, yet we rarely do that with ourselves. Our own inner voice is the most inhuman and sadistic one we will ever hear.

When I made the commitment to living a sober life, I began allowing myself to feel, accept, and not judge any of my emotions. I learned to love myself, despite my demons, and to be present to any anger, fear, or resentment I might have felt toward myself. This approach allowed these emotions to flow through me and disappear.

The next time you experience any emotion that starts to control you and propel you into a downward spiral, become present with it. Let go of any judgments you have about it, including labeling it good or bad. Allow yourself to fully feel the emotion so you understand why it is there. You will find it will soon pass you by.

Seek Out More Useful References

Your referential brain is always taking in information from everything around you and using it to compare one thing to another. Most of this is occurring without your knowledge. Now it's time to take control of the references you want to have in your arsenal.

In the training exercise in chapter 1, you found examples in your life for when you pushed through your fears. Use them. Your own stories can be your most powerful references to shape your future. Take Alice for example. The unbreakable warrior spirit she forged as a result of thriving despite her horrific experience became a reference which she could look back on to handle anything life threw her way. But she had to choose to see it that way. Alice could have forgotten all about those events and buried them. Instead, she chose to integrate them into her being to use as a reference for future growth.

Since we are surrounded by so much information, it is also important to practice consciously choosing what information you want to embed into your mind as a reference. This doesn't mean you can control everything that plasters itself into your subconscious mind. Watching an ad on TV for a car commercial might still affect the car you buy months later. Nonetheless, the more aware you become, the less impact the onslaught of external information will have on your references and ultimately your actions. Awareness will enable you to leverage your references effectively, so you can find empowering meanings to the events in your life and control the emotional impact they have on you, thus enhancing your emotional intelligence.

All the stories, research, and examples I share with you in this book are provided to grow your armory of references. The bigger your arsenal, the more options you have to choose from in making future decisions. As Tony Robbins said, "A larger number and greater quality of references enables us to more effectively evaluate what things mean and what we can do."

To continue finding more useful references, take on new life experiences. For example, act on the list of Courage Transference Actions you wrote down in chapter 1, read new books, watch new documentaries, or meet new people. Do anything that helps you expand your perspective on the world, develop a stronger self-identity, and increase confidence in your own abilities.

Meditate

In one study, Dr. Andrew Newberg taught a group of elderly people experiencing memory problems a short meditation. They were told to practice this for twelve minutes a day for eight weeks. At the end of the eight weeks, Dr. Newburg found "some very significant and profound changes in their brain just at rest, particularly in the areas of the brain that help us to focus our mind and to focus our attention." Participants in the study even reported better memory and an increased ability to think more clearly. Dr. Newberg also discovered the brains of praying nuns, chanting Sikhs, and meditating Buddhists showed increased activity in those same areas of the human brain, as compared to the brains of non-meditators.

It is no surprise then that most successful people today practice some form of meditation. "Meditation more than anything in my life was the biggest ingredient of whatever success I've had," said billionaire Ray Dalio, founder of Bridgewater Associates. A countless number of other millionaires and billionaires echo that sentiment.

Meditation improves your ability to observe your emotions without getting taken over by them. This strengthens the pathway between your animal brain and human brain so you can choose when you want each one to be in charge. There are many ways to meditate and almost all of them produce the same result. Don't let the hundreds of options confuse you to the point that you simply don't do it. Choose one and stick to it.

I do a simple practice every morning and evening. I set a timer for twelve minutes, close my eyes, cross one palm over the other, and sit in silence, focusing on my breath flowing in and out of me. I take long, deep breaths and use that rhythm as an anchor to latch my mind onto. Following this practice doesn't mean thoughts don't show up. When they do, I don't resist them or fight them. I notice them and let them pass, returning my concentration to my breath. The more you practice meditation, the better you will become at finding moments where the mind is in a state of pure stillness.

Chapter 4

BECOMING SUPERHUMAN

A fter waking up from the thought of taking my own life, my mind reeled in a state of chaos. My body, drained of all energy, seemed unable to move. My spirit was nowhere to be found. I stumbled up the stairs into my bedroom, but couldn't stand the sight of my wife. I felt too ashamed to look at her. She reminded me that this was not the person I believed myself to be. From deep inside the pit of hell, I rose up and said to myself, never again. I toppled onto my bed and shut it all out. Tomorrow would be a new beginning.

Less than a week later, I took a nose dive right back into the pit. This time I brought my wife into it with me. I told her I felt like I didn't deserve to live. It should have been me that died in Iraq. I couldn't handle the monotony and boredom of civilian life. I craved a high I believed I could only get by living on the edge of life and death.

Disgusted with myself, I told her that for a brief moment just a few days earlier, I thought about ending it all. With my confession came a flood of tears. This cycle of binge drinking to sobering up with enthusiasm, only to fall back down into an endless number of shots, would destroy me if nothing changed.

I could not believe I succumbed to alcohol less than a week after considering suicide and resolving to change. That blow hit me hard. This time, I had to do something different.

I began studying neuroscience, psychology, and spirituality. I read book after book and learned from every expert I could find in all of these fields. By leveraging the three principles you are about to learn, I learned how to heal my brain.

The Simple Truth of Behavior Change

"Adult behavioral change is the most difficult thing for sentient human beings to accomplish."

—Marshall Goldsmith

What changes are you trying to make? Perhaps it's cutting back on alcohol or losing weight or quitting your job. Have you been able to make those changes? How long have you been struggling with the desire to change, but the inability to do so?

Don't be so hard on yourself if you don't like the answers to those questions. Remember, a lot of that isn't your fault; it's your brain's. It doesn't matter how long you have been struggling to change, once you learn how your brain works, all you have to do is keep putting one foot in front of the other.

"Every brain is wired differently," writes Dr. John Medina. "We have the neural equivalents of large interstate freeways, turnpikes, and state highways. These big trunks are the same from one person to the next, functioning in yours about the same way they function in mine. It's when you get to the smaller routes—the brain's equivalent of residential streets, one lanes and dirt roads—that individual patterns begin to show up. In no two people are they identical."

This book will give you the map for what the highways look like. That is the big picture, illustrating the similarities between your brain and mine. Things will start to look different once we zoom in to those smaller streets. As you read this passage, your brain is being wired in its own unique way. The manner in which you interpret this information, and ultimately the actions you take, will be different from anyone else on the planet.

This phenomenon is known as experience-dependent wiring. A scientist named Quian Quiroga conducted a fascinating experiment illustrating how this works. To save the lives of his epileptic patients, Dr. Quiroga implanted electrodes in various parts of their brains to figure out where the seizures began. He then monitored the electrical activity while showing them a series of images. He was shocked to find that in one patient, a single neuron fired in response to seven different pictures of Jennifer Aniston. In other patients, individual neurons fired in response to pictures of Halle Berry, Bill Clinton, the Simpsons, or the Beatles.

"Our brains are so sensitive to external inputs that their physical wiring depends upon the culture in which they find themselves," writes Dr. Medina. If you watch the TV show *Friends*, but don't watch *The Simpsons*, your neurons will fire when you see a picture of Jennifer Aniston, but not when you see a picture of Homer Simpson. Think about what this means. Everything you have done in your past, everything you do now, and everything that occurs around you every moment of the day affects the wiring in your brain. Who we are evolved from what our environment molded us into.

But even when two people grow up in the same environment, their brains will never be the same. The perception of what shapes reality is dependent on a variety of factors, from our beliefs, to the unique genetically coded wiring in our brain, to our life experience, to anything else that affects our existence. Each of us approaches the world from a distinct perspective.

What this also means is that even if two people appear to have the same problem, they will not have the same solution. While conducting brain scans of thousands of people, neuroscientist Dr. Daniel Amen found "two patients diagnosed with major depression that had virtually the same symptoms yet radically different brains . . . [therefore] treatment needs to be tailored to individual brains, not clusters of symptoms."

There can be no one-size-fits-all formula for transformation. We all need a unique map to guide us through the back roads of our brain. You create this map by walking the streets, by taking wrong turns, adjusting, and finding a new road. Most of all, you keep walking. There is no other way. The secret to behavior change is simple: take action, make mistakes, adapt, and take new action. The greatest lessons are always in the doing.

After recovering from the brink of suicide, I taught myself how to moderate my drinking. That seemed to work for a few years. I built a successful business, ran ultramarathons, and had a very happy married life. Eventually, I realized I was still walking down the wrong path in terms of the person I believed myself to be. Once again, I made an adjustment. After seventeen years of being involved in an on-and-off relationship with alcohol, I quit drinking overnight. Staying sober in situations that ordinarily involved drinking meant entering new territory. Doing something different is always scary. I only realized the freedom I got from my decision to quit drinking by embracing that fear and taking action. No one else could have taught me that. I needed to take the terrifying step into the unknown to experience the bliss first hand.

Right or wrong, good or bad, that was *my* path, and I have no regrets for walking it. I have become the person I am today as a result of all the mistakes I made, along with all the successes I achieved. They have collectively taught me life isn't about finding who we are. It is about creating who we want to be. We get to make that decision every day, through every action we take.

That is the good news. The overarching system—the highways—are the same for all of us, so we all have the same tools to work with. That makes the process of change far less complicated. What makes it fun is that we each get to be pioneering explorers in our own back roads. The bliss of Fearvana lies waiting for us in that pilgrimage. How boring would life be if each of us functioned like robots with identical operations manuals? Or if we knew exactly what every day would look like for the rest of our lives? Life is exciting precisely because there are more than enough unchartered territories to discover until the day we die.

Embarking upon that voyage means being willing to let go of who you are now to become who you want to be. Like the rest of us, you have gotten very comfortable with who you are now, so I know this can be terrifying. Changing your life definitely won't be easy, but it is ALWAYS possible. "The single most important lesson my colleagues and I have learned is that you can literally change people's brains," said Dr. Amen. "You are not stuck with the brain you have. You can make it better."

Change seems impossible at times only because we haven't become aware of things like the myth of free will, the value of fear, and the beauty of adversity. The journey won't be easy, but it is simple because, ultimately, there are only two things we can change.

The Only Two Things We Can Change

"When we are no longer able to change a situation, we are challenged to change ourselves."

—Victor Frankl

In Jack Canfield's bestselling book *The Success Principles: How to Get from Where You Are to Where You Want to Be*, he states, "If you want to be successful, you have to take 100 percent responsibility for everything that you experience in your life. This includes the level of your achievements, the results you produce, the quality of your relationships, the state of your health and physical fitness, your income, your debts, your feelings—everything!" Yet, so many of us surrender to the circumstances of our lives. Some people blame external forces, such as the economy, the president, or their job. Others blame internal forces, like depression, anxiety, or addiction.

Accepting the lack of control we have over our brain is not meant to absolve us of responsibility for our own transformation or resign ourselves to the hand we are dealt. On the contrary, this acceptance is a stepping stone to taking responsibility for reclaiming control of our consciousness and ultimately our destiny.

Only by accepting she had no control over the fear showing up in her brain while climbing up Bell Rock was my wife able to take responsibility for that fear. She then took control, instead of letting the series of second darts consume her without her awareness.

Moving from a place of awareness and acceptance about the true nature of free will, there are then only two things for which you can take responsibility:

your actions and your attitude. If you don't like your external reality, you can change your behavior to create a new reality, or you can change your attitude in response to your current reality. Either of those changes will lead to a better quality of life, but both require control over consciousness. Practice constantly asking yourself, how can I be responsible for this? Using the human brain to become more aware of your circumstances will give you the means to take charge of how you respond to those circumstances.

Imprisoned in a concentration camp in Auschwitz, Victor Frankl had minimal control over his external reality. The guards who held him prisoner restricted his actions through brute force and torture. Without the ability to change his behavior, he could only control his attitude about the world around him. By seizing control over his mind, he found value in his suffering and transformed the agony of life in a concentration camp into a meaningful experience.

On the other side of the coin, there is my friend Forrest Willett. He is an outstanding example of someone who altered his behavior to improve his experience of life. Forrest lived the kind of life most people dream of. He had seven different businesses, three homes, a lovely wife, and a healthy two-year-old son. Then everything went to hell. At the age of thirty-one, Forrest got into a severe car accident that left him with a debilitating brain injury.

His doctors and therapists told him he would never be able to live a normal life again. Without any hope of recovery, Forrest laid in bed for five years, overcome with depression. One day he saw Jack Canfield on TV, discussing his book *The Success Principles: How to Get from Where You Are to Where You Want to Be*. Those words ignited a spark that transformed Forrest's life.

He stopped blaming the world around him and feeling sorry for himself. He began taking complete responsibility for his fate. Although his speech therapist told him it would be impossible to read a four hundred-page book, Forrest bought it anyway. Through endless hours of focus, commitment, and laborious hard work, Forrest not only taught himself how to read again, but to speak and write as well. Today Forrest is a successful public speaker and published author who uses his story to help others. His book *Baseballs Don't Bounce* documents his incredible journey in his own words.

Forrest not only changed his external world, but he changed the internal structure of his brain on a neurological and physiological level as well. Like Forrest, you too have the power to change your brain.

The One Superpower We Are All Born With

Our superpower is called neuroplasticity, which is defined as the brain's ability to reorganize itself by forming new neural connections throughout life or, in simpler terms, the brain's ability to change itself.

As you read this book, the knowledge is being relayed to all parts of your brain through neurons and the synapses that connect them. Think of the synapses as little roads that transport information from one neuron to the next to form large networks throughout your brain. The reason you can get on your bike and pedal it down the street with such ease is because your brain has created pathways that tell your body how to respond to the stimulus of being on a bike. When one neuron fires, it sets off a sequence of neuronal firing that creates your thoughts and actions.

This neuronal firing and your resulting thoughts are part of a two-way street. "Mental activity stimulates brain firing as much as brain firing creates mental activity," says Dr. Daniel Siegel. This means you are not at the mercy of the patterns that occur in your brain; you can direct your attention to change those patterns.

These two methods for changing your brain are known as top-down and bottom-up neuroplasticity. I have separated our brain into two parts for the sake of simplicity and understanding, but they do, of course, work with each other on a second-by-second basis. The animal brain affects the human brain and vice versa. As Csikszentmihalyi states, "Attention shapes the self, and is in turn shaped by it."

Bottom-up neuroplasticity is when the cognitive biases in your subconscious animal brain shapes the entirety of your brain without your awareness. This is occurring right now as you read this book. Bottom-up neuroplasticity is essential to learning patterns that allow us to respond rapidly and effectively to our environment. But to realize our greatness, we must separate ourselves from being at the mercy of our external world.

The ability for consciously directed change comes from the top down when we use attention, focus, and awareness to mold our brain. That is the magic of neuroplasticity. By altering our thought patterns, behaviors, and emotions, we have the capacity to build new roads in our brain, which leads to new actions. Using the gift of neuroplasticity, Forrest was able to change his brain, despite everyone telling him it was impossible. No matter how old we are, each of us has this power to reconstruct every part of our brain on a physiological level.

As Dr. Norman Doidge writes in *The Brain That Changes Itself*, "Neuroplasticity is neither ghettoized within certain departments in the brain nor confined to the sensory, motor, and cognitive processing areas." The hypothalamus, amygdala, and hippocampus are all plastic.

Dr. Micheal Merzenich, often known as the world's leading researcher on brain plasticity, puts it this way: "You cannot have plasticity in isolation. It's an absolute impossibility." Whatever state you are in now, you can change your behavior to create the life you want to live—financially, physically, spiritually, and mentally—by using the three principles of neuroplasticity.

The Three Principles of Neuroplasticity

Hebb's Law

Hebb's Law states that neurons that fire together, wire together. This is the most fundamental concept of neuroplasticity. Imagine riding a sled down a snow-covered hill. After a few exhilarating rides, your sled creates a deep track in the snow. When you push off at the top of the hill again, your sled will find those ruts and follow them whether you want it to or not. Similarly, when separate parts of your brain are activated together on a recurring basis, they become locked together so that one always functions in conjunction with the other. This is how habits form.

In my case, stress and alcohol had fired together so often that they became "wired" together. The problem wasn't the stress or even that my brain led me to perceive alcohol as the only response to stress; the real problem was that I focused my attention on the alcohol after the thought of it entered my brain. That allowed the quantum Zeno effect to take place.

Quantum Zeno Effect

Patterns only form in your brain when neurons are activated long enough for Hebb's Law to take effect. The quantum Zeno effect is responsible for this prolonged activation. "The more attention the brain pays to a given stimulus, the more elaborately the information will be encoded," writes Dr. Medina in his book *Brain Rules*.

Using the metaphor above, you can think of the quantum Zeno effect as your weight on the sled. If you merely push an empty sled down a hill, it won't make much of an impact in the snow. Your weight creates the deep track for that sled to ride on. Generating the metaphorical weight on the sled requires conscious activation.

Forrest spent hours, days, and weeks focusing on nothing but the book in front of him. By directing his attention for long periods of time, the quantum Zeno effect rewired his brain, giving him the ability to read, write, and speak again. The more we train ourselves to step outside of the animal brain and direct the focus of our human brain, the greater our ability to consciously generate new mental patterns that lead to new behaviors.

Use It or Lose It

A brutal war is waged every moment of every day inside your brain. Like many wars, this internal war is fought for land; more specifically, for space within your brain. Within the animal brain and the human brain, there are numerous "land masses," each responsible for various functions that cumulatively make up the human experience. The neurons that live on each of these land masses battle each other for space. This battle illustrates the principle of "use it or lose it."

Dr. Merzenich was responsible for shedding light on this principle to the world. In a revolutionary experiment that transformed neuroplasticity from myth to fact for many nonbelievers, Merzenich spent a countless number of months mapping a monkey's hand. The tedious procedure involved inserting a microelectrode next to a neuron within a small section of the monkey's sensory cortex. This was followed by a series of taps on the monkey's hand until he found which part of the hand activated that particular neuron.

Through repeated surgeries and insertions, Merzenich and his team figured out the millimeter-sized land mass associated with every inch of the monkey's hand. So when he touched the tip of the monkey's pinky finger, he saw a different area of the brain light up than when he touched the center of the palm. After mapping the entire hand, Merzenich amputated the monkey's middle finger. A few months later, he remapped the hand and discovered that the land mass associated with the middle finger had been conquered by the land mass associated with the neighboring fingers. So when Merzenich touched the monkey's other fingers, it activated the same section of the brain that was formerly activated by the middle finger.

Despite how prevalent the myth is that we use just 10 percent of our brain, the truth is that we use all of it. As Merzenich demonstrated, if we don't use a part of it, the brain will assign a new use for that part. Numerous other experiments conducted after Merzenich's have further validated the principle of "use it or lose it," otherwise known as competitive plasticity.

The brain is a muscle like any other muscle in your body. If you don't work it out, if you don't use the parts of it necessary for your success, they simply die out in the never-ending war for cortical space. Neurons that do not consistently fire together inevitably cannot wire together. So if you don't keep practicing the behaviors you want, your brain assumes you no longer need those neurological highways and kills them off to build other ones. If you spend every day watching four hours of television as opposed to reading a book on personal development or doing anything else to drive your life forward, you can literally destroy your brain.

In their research, professors Mellanie Springer and Cheryl Grady discovered that the more education a person has, the more he or she uses the frontal or temporal lobes when performing various cognitive tasks. The greater the level of education, the greater the activity in the human brain. Education acts as a brain workout that builds our neuronal muscles.

Working in conjunction with Hebb's Law, the "use it or lose it" principle ensures that the active parts of your brain will defeat and conquer the unused, inactive parts. The more you repeat the same pattern, the more armies you place in that land mass and the stronger that behavior becomes.

Through your actions, you get to decide who wins the battles in your brain. The exercise below will help you ensure your greatest self emerges victorious in this never-ending war.

Training Exercise

The following five-step formula is one of the most powerful tools I have created to help you change your brain. It has helped people with PTSD, anxiety disorder, addiction, depression, daily stress, fear, food cravings, and procrastination, among other things. I am confident that if you consistently use this, it will work for you, no matter what the situation or emotion in your way. I will demonstrate how it works with an example from a client of mine for each step listed below. There are, however, three caveats for this to work:

1. You must want to put in the work and be willing to make changes at this point in time. This sounds obvious, but you will be surprised how often people say they want to change, but they don't truly want to change. Or they don't want to change right now; they want to wait for another few months. To ensure you are ready to make changes, I recommend asking yourself the question represented by the acronym AIWATT. I learned this question from *New York Times* bestselling author and leadership coach Marshall Goldsmith. It goes like this: "Am I willing, at this time, to make the investment required to make a positive difference on this topic?" Are you? If so, then move on to the second caveat.

2. You must take full responsibility for the event, without blaming anything else outside of you. Remember you can only change two things: your attitude or your behavior. As long as you accept the responsibility for your growth lies in your own hands, this exercise will help you change one or both of those.

3. You actually have to go through the steps and do the exercise for it to work. This might also sound obvious, but, once again, you would be surprised by how often people who have used this exercise in one context forget to use it in another. Then they wonder why they are unable to change the behavior.

Are you willing to accept these three caveats? Assuming you are, then here is the "Unstoppable Warrior Formula." To make it easy for you to remember, so you will actually use it, it is spelled out as the LMNOP cycle.

Step 1: Label and Language (L)

Any time you find yourself stuck in any way, label the emotion you are experiencing to release yourself from the impact of it. Then immediately change your body language into a "power pose."

Dr. Matthew Lieberman, a psychology professor at UCLA, has shown that labeling an emotion reduces activity in parts of the animal brain related to fear and emotions. Simultaneously, it increases activity in parts of the human brain associated with focus and processing emotions. Without being at the mercy of our emotions, we can consciously choose where we direct our actions.

Social psychologist Amy Cuddy found that a "power pose" reduces the stress hormone cortisol and increases testosterone in your body. Just as our minds control our bodies, the system works the other way as well. To assume the "power pose," stand tall, open your body into an expansive position, and act as if you are the most confident person in the world. Temporarily faking it will help you actually feel more confident to take the action you need to take in step five.

Example

In chapter 1, I briefly mentioned my client Bob who suffered from severe anxiety every time he got on his computer to write. His physiological reactions were so intense that every time he attempted to work, he could not handle it and watched TV instead. This exercise helped him solve his problem.

First he labeled his emotion as anxiety and kept doing it for as long as he needed to let the troubling effects subside. By being fully present to the emotion, he was able to separate himself from it and process it, which caused his physiological reactions to die down. Next he worked on his body language. Every time I saw Bob on Skype, he was slouched over in his chair. While labeling the emotion, he immediately reversed his posture and sat upright, adopting a pose of power and confidence.

Step 2: Meaning (M)

The next step is to ask yourself about the meaning you have attached to the event, the emotion, or to both. Emotions and events have no inherent meaning. The meanings we assign will shape our experience of life. As we learned from Professor Gazzaniga in the last chapter, our brain is always finding meaning. In this step, we are digging into our brain to find out what that meaning is.

Example

By bringing awareness to the meaning behind his anxiety, Bob figured out he had created a meaning in his subconscious mind that no one would want to read his writing. He thought everyone would think it was worthless. He also created a meaning for the anxiety itself. He believed the anxiety meant he was weak and pathetic. This caused a rapid descent into second dart syndrome.

Step 3: It's Not You; It's Your Brain (N)

We often define ourselves by the emotions we are experiencing. In this step, you are acknowledging to yourself that this is just your brain stuck in a pattern outside the realm of your awareness. This further separates you from your uncontrollable subconscious forces. All you have to do is tell yourself, "This is not me; this is my brain, beyond my control." Do this in any form that fits your conversational style. The point is to remind yourself that whatever is occurring right now is not you. IT IS NOT WHO YOU ARE.

Example

Bob simply told himself, "This is not me; this is just my brain experiencing a feeling of anxiety that is beyond my control."

Step 4: Opt for the More Empowering Meaning (O)

Now that you have separated yourself from your animal brain, it is time to activate your human brain to ascribe a new meaning either to the emotion, the event, or both.

Whatever the emotion may be, don't resist it. Welcome it, and find the positive motivation behind it. Every emotion shows up for a reason. Your body, mind, and spirit have your best interest at heart; trust them and dig deep to find that empowering meaning behind the emotion.

For the event, reframe it in your mind and search for references that allow you to create a new meaning to it. An effective way to do this is to step into the future, then look back on the event and ask yourself, "How could I be grateful for this event?" Expressing gratitude will help you find a more empowering meaning.

Deep down, your subconscious brain might not believe the new meaning you choose because it is still stuck in its old patterns. Opt to focus on the meaning anyway. This, combined with the next step, will start to create new neurological pathways in your brain.

For even more impact, focus on a meaning based on something or someone greater than yourself. This releases oxytocin in your brain, which has been shown to enhance our ability to act in the face of fear and stress.

Example

Earlier in his life, Bob had been extremely successful. He chose to use his past successes as a reference to create a new meaning, that people would want to hear what he has to say. He came to realize others would value his insight and his book could serve a purpose.

He also created an empowering meaning for his anxiety. He made it mean that the only reason he even felt anxious is because he cared enough about people to want to help them with his writing. This meaning allowed him to leverage the energy of the anxiety and take it with him into the final step.

Step 5: Purpose and Preemptive Strikes (P)

This final step is where the quantum Zeno effect does its job and rewires the brain through Hebb's Law. There are two parts to this step:

1. Taking an action aligned with your true purpose
2. Continue ingraining the new behavior with preemptive strikes.

Now that you have chosen a new meaning for the event and/or the emotion, use the energy of that to take a new action, something more in line with the person you want to be. You MUST do something different than before, no matter how small, to build new brain patterns. The action can be a physical one or a mental one; it depends on the pattern you are trying to break.

For example, when I first became sober, I found myself stuck in the same mental pattern of hating myself every time I didn't do something I "should" have done, like write for an hour. I would then get stressed to the point of feeling overwhelmed over how much work I needed to do. Using the LMNOP cycle, I learned to shift my mental patterns to loving myself, forgiving myself, and allowing myself to focus on making progress, instead of reaching the perfection of a destination. By directing my attention to something other than the old pattern I was stuck in, I created new neural pathways in my brain. On the other hand, as you will see in Bob's example below, the action you take can also be a physical one.

Another key factor to consider is the manner in which you make the transition from your old behavior to the new one. You can either smoothly flow into a new action or harshly interrupt the previous pattern to move on to the next one.

In my example above, I found it more valuable to flow gently and lovingly into a new thought pattern. However, when I find myself too lazy to work out, I need to interrupt that pattern harshly. I berate myself for being a fat, worthless, lazy scumbag. I go Marine Corps drill instructor on myself to get me charged up. I am not saying that is what you should do. It might sound a little ridiculous, but it does get me to the gym.

What ultimately matters is that you perform an action in line with your true purpose. Experiment and have fun to see which method better enables you to take action. Don't take the process so seriously. Play around with it because there is no right or wrong way to do things; there is only what works and what doesn't.

To then prevent yourself from getting stuck in that same pattern in the future, use preemptive strikes. A preemptive strike is to set a clear, detailed plan for how, when, and what you will do the next time you know you will be in that disempowering pattern again.

Psychologists Peter Gollwitzer and Veronika Brand-Statter have found that by creating a strategy for how you will engage in a very specific action ahead of time, you triple your chance for success. Elderly patients in Scotland who used preemptive strikes after going through knee and hip replacement surgery recovered almost three times faster than those who did not. They did so by writing down exactly when and how they would take baths or take walks or engage in any behavior that brought them pain. This allowed them to better navigate the fear and stress of that pain.

You also can use preemptive strikes to prevent any obstacle you know will show up. For example, one day I woke up after only a few hours of sleep. I was exhausted and had two client calls scheduled that morning, followed by a five mile run. I knew that after those calls, I would want to take a nap.

To combat the inevitable fatigue, I put on my running shorts, shoes, and GPS running watch. I then sat near the front door for both my calls. As soon as I got off the second call, all I had to do was step out the door. By preemptively preparing for an obstacle I knew would show up, I made it as psychologically easy as possible to take the proper action.

To use preemptive strikes, you can create a written plan, like the elderly patients in Scotland, or just think of the plan as I did to get my run in. I myself use written plans for organizing my day and running my business, but in that one particular scenario, I didn't need to.

Ultimately, the goal in this final step is to prepare for the next time you might need to use the LMNOP cycle so you don't waste cognitive energy. With preemptive strikes, all you have to do is follow the plan without thinking.

Example

Bob would interrupt his pattern by sitting up straight, smiling, pumping his fists, anything that energized him. He could then write for five to ten minutes. Gradually, he increased the time, but initially he just needed to create a new pathway. He then wrote down the strategy that worked for him and used it as a preemptive strike for the next time he needed to sit down and write.

Deeply ingrained patterns will not change overnight, but by using the LMNOP cycle regularly, in time, you will be able to create new pathways that unleash the legend that lives within you.

Chapter 5
TIME TRAVELING TO CHANGE YOUR PAST

I once worked with a woman named Rachel who was at a transition point in her life. She had just made the decision to abandon a lifestyle of clubbing and partying. Instead, she committed herself to a daily spiritual practice, eating only raw foods, and other positive changes. Despite her commitment, she found herself resisting the change and could not understand why. She knew this new lifestyle was exactly what she wanted, yet there was a deep-rooted fear holding her back.

Buried within her implicit memory lay the answer to her troubles. Working together, we wandered into her past to visualize the events that took place when she chose this new way of life. She remembered her old friends abandoning her because they could not tolerate her "extreme" raw food lifestyle. A series of these experiences led her to believe the raw food lifestyle meant having no friends. Inevitably, being a social creature who craves human connection, she found herself scared to embrace that way of life. By bringing that implicit memory into her awareness, she was then able to shift the meaning she created for those events.

To help Rachel choose an empowering meaning, I guided her to visualize the healthier future she wanted to live in. When I sensed her desired future had been validated on an emotional level, I verbally steered her back into the past. At first, it was a shock for her to go from a place of joy to a place of darkness. But from that place of joy, she saw the experience in a different light.

Suddenly, in what can only be classified as a light-bulb moment, she realized her friends hadn't abandoned her; they gave her the space to live her purpose. "They gave me the wings to fly," she ecstatically realized. By changing the meaning ascribed to those past experiences, she no longer saw them as negative moments in her life. For her, the past had been altered. The "reality" itself changed because perception shapes reality, and perception is based on memories that constantly lie to us.

How Your Brain Turns an Experience into a Memory

"Every special date and anniversary, every advertisement, every therapy session, every day in school is an effort to create or modify memory."
—Dr. Joseph Ledoux

Memories create the series of habits, associations, and patterns that make you who you are today. Like it or not, you are a product of your past. Your present is molded by it, and your future is dependent on it. But the past doesn't have to define you. It isn't real. Your memories are as plastic as your brain. With a proper understanding of memory, you can manipulate it to let go of your past, push through your fears, and experience the bliss of Fearvana. There are three steps involved in the process of converting an experience into a memory:

1. Acquisition: Acquisition occurs when your brain, working in conjunction with the bodyguard at its base, receives and processes external stimuli. As you read this paragraph, your brain is starting to form a neural network based on the information it's acquiring.

2. Consolidation: Most of the information that comes into your brain is lost in short-term memory, but some of it becomes a part of your long-term memory. To implant an experience into your long-term memory, neurons are connected through pathways that collectively form large neural networks. These are physical maps that materialize as structures in your brain representing a memory. The process of strengthening these pathways to build the construction of a memory network is called consolidation. How each memory is consolidated depends on various factors, including how much attention you paid to the event, the emotional impact it had, and the number of senses it engaged. If you simply skim this book, it will fade away from your memory. If you focus on the content and apply it, you will gain the experiential memory required to consolidate the knowledge into your subconscious.

3. Retrieval and Reconsolidation: Retrieval is when you pull a past experience from your brain and bring it into the present. During retrieval, your brain activates a neuron that triggers the other neurons in that particular memory network. If a part of a memory is activated, such as the sights, sounds, or tastes you experienced, it lights up the rest of the neurons in that network. This is why the song "When You Say Nothing at All" triggers the memory of my ex-girlfriend. Reconsolidation is what occurs during retrieval. It is your brain drawing information from various regions and putting these pieces together as a consolidated memory to bring into your consciousness.

The efficiency of each of these steps is dependent on many factors, including genes, health, stress levels, and belief systems, to name a few. Regardless of where you are now, though, your memory is plastic, so it can be improved.

There are two kinds of memories you have the power to mold: implicit and explicit memory. What did you do yesterday evening? To answer that question, your brain activated the neural network of yesterday's events and retrieved that memory map to tell the story of what you did. You actively brought the past into your present awareness. The conscious direction of your mind into your past is known as explicit memory.

On the other hand, if you were to put down this book, step outside, and get in your car, assuming you know how to drive, would you have to think about it? The reason you can drive with such ease, or walk through your home, or even know how to walk for that matter, is because of implicit memory. Implicit memory runs on autopilot without your human brain.

When you entered into this world as a helpless infant, these memories were responsible for your transformation into adulthood. In fact, researchers believe that in the first year and a half of our lives, we only encode memories implicitly. According to Dr. Daniel Siegel, the three features of implicit memory are, as follows:

1. You don't need to use focal, conscious attention for the creation of implicit memory.
2. When an implicit memory emerges from storage, you do not have the sensation that something is being recalled from the past. (You don't think about the first time you learned how to walk every time you walk.)
3. Implicit memory does not require the participation of the hippocampus (the human brain's role in memory).

Your implicit memories are responsible for your beliefs, your subconscious mental models, your sense of right or wrong, and the triggers that cause you fear, stress, and anxiety.

What If You Could Be Fearless?

"Memories influence every action and pattern of action you undertake."
—Dr. Daniel Amen

The neural networks that form a memory live in many different areas of your brain, but there are two areas most active in the creation and storage of memory: the amygdala and the hippocampus. The amygdala is responsible for implicit memory, and the hippocampus is responsible for explicit memory. That is an oversimplification, but it is a useful one in helping you understand the two kinds of memory.

Dr. Siegel calls the hippocampus "the master puzzle piece assembler." It compiles the information it receives from multiple areas of your brain to produce memories, meanings, and emotions for any event. It also helps consolidate the information stored in short-term memory, turning it into a long-term memory you can recall in the future. When I asked you what you did yesterday, those events were probably not at the top of your mind. By answering the question, as Dr. Siegel states, your hippocampus "literally link[ed] together the neurally distributed puzzle pieces of implicit memory."

The conscious activation of your memory turns the implicit into explicit. Various parts of the brain work together to form these implicit memories, such as the basal ganglia, which is "the habit center" of your brain, but the amygdala is primarily responsible for this task. The amygdala, or fear center of the brain, stores emotionally charged memories to help you avoid future danger. If you are like me and you were a bitten by a dog as a child, your amygdala imprinted that experience into your implicit memory, possibly causing you to have a natural aversion to dogs, or at least the kind of dog that bit you. This is what makes the amygdala the central player in the creation of all learned fears. Wouldn't it be great if we could just get rid of it? Not really.

In 2010, researchers Justin Feinstein and his colleagues discovered a woman, whom we shall call Mary, with an extremely rare disease that left her without an amygdala. She was a goldmine for neuroscientists. They did everything they could to scare this woman, but nothing worked. They made her watch scary movies; they gave her snakes; they put spiders in her hand; but none of them registered a fear response in her brain. Imagine being completely fearless. You could quit that job you hate, run that marathon, write that book, start that business, and do all those things you have always wanted to do but have been held back by fear.

Turns out, it's not that easy, as Mary discovered. One night while strolling through a park alone, Mary was attacked by someone wielding a knife. What would you do if that happened to you? More than likely, you would not return to that same park alone and in the dark, at least not in the next week. The reason you would stay away is because you have a functioning amygdala that remembers the danger. Mary did not have this capacity. The very next night, she went back to the same park, once again, alone.

Mary had been attacked at gunpoint, had her life threatened, and was almost killed in a domestic violence incident—all because she had no amygdala to process fear and keep her out of life-threatening situations. The amygdala helps keep us alive by learning what to fear.

"You don't learn how to be afraid; your amygdala doesn't have to learn what to do; it learns what to do it in response to [stimuli]. So it learns what stimuli it should respond to," says Ledoux. "So it's learning and memory in that sense that we call an implicit kind of memory where you don't have to have any conscious involvement." Remember, no matter what fears show up, they are not bad or unreasonable. You don't control their existence; the amygdala does.

Without the activation of your conscious self, how can you be held responsible for your fears? Your implicit memory has implanted them into you. This is why so many of us are held captive by the events of the past and the fears they have created within us. The good news is that these memories are not actually true, and they most definitely are not set in stone.

Your Past Is a Lie

"Until you make the unconscious conscious, it will direct your life and you will call it fate."

—**Carl Jung**

Shortly after the tragedy at the Twin Towers on 9/11, psychologists conducted a survey with hundreds of people about their memories of the event. In a follow-up survey one year later, they found that 37 percent of the details were different. Within three years, that number rose to almost 50 percent. Some of the memory alterations were minuscule; others involved an entire shift in the story line. Some people even remembered being at a different location that morning.

After the study, Elizabeth Phelps, one of the lead researchers, wrote, "What's most troubling, of course, is that these people have no idea their memories have changed this much. The strength of the emotion makes them convinced it's all true, even when it's clearly not."

In another study, conducted on three hundred people convicted of crimes in the United States who were later proven innocent through DNA testing, researchers found 75 percent of them were sent to prison based on false memories of eyewitness. The eyewitnesses did not know they were lying; they simply believed their memories to be facts. Truth is nothing more than what we believe it to be, and those beliefs are as malleable as the memories that created them.

To demonstrate how our memories can be manipulated, psychologist Elizabeth Loftus has repeatedly proven that as many as 50 percent of the people in any given study could be tricked into believing a fabricated event. Using what she calls a "false feedback" technique, her team embedded into the minds of their research participants fake memories, such as finding yourself lost and crying in a shopping mall as a child, almost drowning before being rescued by a lifeguard, and getting attacked by an animal, to name a few.

In numerous other studies with people from all walks of life, even those trained to handle stressful situations like those in the US Special Forces, Loftus has demonstrated the ease with which false memories can be implanted into the human brain. "Memory works a little bit more like a Wikipedia page: You can go in there and change it, but so can other people," she says.

Most of us think memory is when we remember a past event. We believe it "works like a video camera, accurately recording the events we see and hear so that we can review and inspect them later," summarized psychologists Dan Simons and Chris Chabris. In reality, memory is like putty; it can be molded by all who possess it. You might have been in a situation where two people recall completely different "facts" about the same event. This occurs because of how a memory is reconsolidated in our brains. Every time we explicitly recall a memory, we are not remembering the event itself, but the last time we remembered that event.

"We learn, we store, we retrieve, and when we retrieve the next time, we are not retrieving the original experience—we are retrieving our last retrieval," says Ledoux. "In other words, upon retrieval, a new memory is formed." In his last line lies the secret to changing your past.

When you consciously go back in time to recall an event, the memory is summoned from your hippocampus, which works with the amygdala and other parts of your brain to remember your past. The act of remembering alters the neural network of that memory, creating an entirely new structure of neural connections. So every time you think about a past event, the "reality" of that event changes, based on your current state of being, your current level of awareness, and the present conditions in which the memory is recalled. Since a memory is formed by your conscious remembrance of it, not by the event itself, altering the conditions in your brain during recall can recreate the neuronal map of your memories and the stories they tell.

If you make yourself happy now and then travel back in time to a sad memory, the joy you feel in the present will change the neurological formation of that sad event. By choosing your present state of being and then going into your past, you can change the effect the past has on you today.

For a while, Rachel's past kept her imprisoned by fear. By first sending her into her desired future, Rachel was able to then travel back in time from an empowered place. Her state of being in the present delivered a wave of positive emotions into the neural network of that memory. Those emotions became the fuel that made her implicit memories explicit, allowing her to consciously change the impact and, in turn, the content of those events.

What this means for you is that your memories might not be true. But don't let this information lead you to questioning every aspect of your past. That could drive anyone insane. Instead, use the malleability of memory to your advantage. Ultimately it doesn't matter how "true" your past is. Memories function the way they do because all that matters is how your past helps drive you forward today. As LeDoux explains, "The brain isn't interested in having a perfect set of memories about the past. Instead, memory comes with a natural updating mechanism, which is how we make sure that the information taking up valuable space inside our head is still useful. That might make our memories less accurate, but it probably also makes them more relevant to the future."

Memories come and go based on the "use it or lose it" principle of neuroplasticity. Self-awareness allows us to choose the kind of information we want to occupy neural real estate. Without it, the battle is lost to the implicit fears and stressors our ancient animal brain thinks it needs to keep us alive. The animal brain might be long overdue for an upgrade from the primitive lifestyle it is still accustomed to, but until that happens, we must activate our human brain to alter the content of memory. Once we make the choice to consciously travel back in time from an empowered state, we must act quickly. We don't have a lot of time to take the next step in changing our past.

In a study at NYU, Dr. Daniela Schiller and her team of researchers showed two groups of participants a random sequence of blue and yellow squares on a screen. On the first day of the experiment, group 1 and group 2 were both occasionally administered an electronic shock when exposed to a blue square. Both groups developed a fear memory associating blue squares with pain. Our "associative memory" is constantly forming associations between the various elements that come together to form a memory, so that when one element is activated, it triggers the other as well.

On the second day, the researchers reminded both groups of their fear by exposing them to a blue square. Inevitably, the association they learned the previous day triggered a fear response in the participants' brain, causing them to sweat at the sight of that dreaded blue square. With the fear memory now activated, the researchers left the room and returned to group 1 after a few hours. They then flashed a sequence of blue and yellow squares without administering a shock. They continued this until the group no longer feared the blue squares.

With group 2, the researchers returned after six hours and repeated the same procedure until they too no longer feared the blue squares. By exposing both groups to the source of their fear and removing the pain associated with it, they eliminated the fear response from automatically showing up in their brain.

On day three, the researchers once again showed both groups a blue square without a shock. This time, only group 1 showed no fear. Group 2 still remembered the association it had formed on day one and reverted back to that old fear response. "It was pretty astonishing," said Dr. Schiller. "It had so many implications for why some therapies only work temporarily. The original idea was that you could forget but you'd always have the original memory stored somewhere. The new theory is that memory can be updated, and there is a window in which this can be done."

The researchers brought the participants back one year later, and their response was the same. Group 1 had permanently eliminated their fear of blue squares, while group 2 still remembered the pain accompanying the flash of a blue square. This six-hour window for altering a fear memory has been found to be present in rats as well.

What this means is that to change the past, we need to activate a memory from an optimistic present state and modify it within six hours, just as Rachel did. Your past helped shape the fears that keep you imprisoned in your present, so altering your memories is often a necessary step to move from fear to Fearvana. Let's work on adapting your history to serve your present self and your future self.

Training Exercise

I want to make it clear that just because we are delving into the past doesn't mean we all need therapy to thrive. The problem with many forms of therapy is that it takes you back in time from a very disempowered state in the present. Simply opening up your heart about the past doesn't free you from it. When done incorrectly, the time traveling process only aggravates the stressors and fears caused by the very event that brought you to therapy in the first place. It reinforces the negative impact of that past event by creating an unpleasant association between the neural network of that memory and your current distressed state.

I experienced this first hand as a veteran diagnosed with PTSD by the Department of Veterans Affairs. Before I taught myself how to heal my brain using everything I am now teaching you in this book, I went to a therapist to overcome my psychological obstacles. I used to walk out of that office more miserable than when I walked in, and very often, I would drive straight to the liquor store.

My therapist told me this was a normal, even necessary, part of the process to heal past wounds. Although I know he genuinely wanted to help, after learning the inner workings of the mind, I came to realize his approach was far from the truth. The methods below allowed me to finally find value in the guilt I felt over losing my friend in Iraq.

Another problem I found with therapy is that traveling back into the past is useless unless we use it to drive us forward. It is pointless to spend years on a therapist's couch, analyzing and interpreting every second of your past, with no clarity as to what you want to gain from it. You don't need to waste all that time.

We only need to delve into the past to the extent that it negatively affects who we are today and who we want to be tomorrow; otherwise, what difference does it make? Those moments are now over. We can't do anything to get them back or alter the actual events. All we can do is change our memory if and when we need to.

Back toward the Future Exercise

In this exercise, our focus will be to travel back in time with the sole intention of aligning the events of our past with the future we want to step into.

Step 1: Choose one of your goals where you are struggling to make progress. Choose something you think you are having a hard time with because of some subconscious force buried in your implicit memory.

Step 2: In chapter 1, you wrote down or created a visual presentation of your accomplishments. Now we are going to build on that exercise to bring those experiences into the forefront of your awareness. If you are very clear on the future life you want to create, you can also visualize what it looks like. More often than not, though, it is harder to imagine yourself in a situation you have never been in, as opposed to recalling one you have personally experienced.

A past event has a neural network that incorporates your senses and your emotions, so it could have more of an impact to bring that into your awareness. But if you have a strong imagination and a proclivity for visualization, as Rachel did, feel free to bring your future self into your awareness as well. Either way works. The key is to ensure something positive is front and center of your awareness and you feel the impact of it. You can even do this by finding activities that bring you joy, like going for a run, and use them as tools to reframe your past.

Remember Alice from chapter 3? She is a perfect example of someone who altered a traumatic past by infusing positivity into her life. "I don't remember a period of more than a few days where it actually completely consumed me," she said of her past. "I remember being able to find joy during those times in something, in my little cousins, in being outside, in movement . . . I'm really playful. I love to be outside, so being around that in other people was a reminder of what was real." She never let her past consume her because she found avenues of joy to construct her desired reality.

Step 3: Once you are focused on something empowering and feeling the favorable emotion deep within you, go back into your past to find the event keeping you from taking action to better your future. You might know what that event is ahead of time. If not, use all the awareness exercises in this book to get clear on where in your history you experienced a moment that created some disempowering belief about yourself or the world around you.

Keep the positive sensation in the foreground and allow the past event to glide behind it. "Have a positive experience be prominent in awareness while the painful one is sensed dimly in the background," writes Dr. Rick Hanson in his book *Buddha's Brain*.

Step 4: This next step needs to be taken within six hours of activating the memory. If you follow along with this entire exercise, that shouldn't be a problem as you will flow seamlessly from the last step into this one. You have two options in this step. Use one or both of them, depending on what works best for you and what you need:

1. From that empowered place, choose a new meaning for the event. For example, although I am not proud of wasting a year and a half of my life with drugs, I now know that lifestyle also led me into the Marines, so I have no regrets. I haven't forgotten the past, but I reframed it. You too probably won't forget what happened to you, especially if it was emotionally charged, but you can let go of the impact that event has on you.

2. From that empowered place, search for examples in your past in direct contrast to the block you are currently experiencing. For example, someone I once worked with felt terrified of traveling to a new place by himself. Fear kept him from taking action to venture out of his home. To combat the fearful memory, he focused his attention on the past

experiences when he mustered up the courage to engage strangers in conversation or wander through an unknown city alone. This exercise didn't eliminate the fear the next time he traveled to a new country, but it gave him confidence to do what was needed to get on that plane.

Step 5: Notice when the past event is holding you back and keep conditioning the new meaning to it and/or replacing it with the other event that contradicts it.

Step 6: Keep a daily record celebrating positive experiences. This can be in the form of a gratitude journal or accomplishment log. I use a variation of both, depending on the events of the day. Because of negativity bias, our mind implicitly tends to focus only on the negative memories. In this step, we are becoming proactive about consciously directing our awareness only toward the positive.

Do whatever it takes to feel the positive emotion so it continuously plants itself into your memory. For example, a few days after permanently sobering up, I finished a run and sat on my lawn listening to Bruce Springsteen's "Paradise." With the sun on my face, my puppy sitting beside me, and the serene sounds of the Boss seeping into my soul, I basked in the bliss of knowing I had finally become the person I always believed myself to be. The beauty of the moment was so intense, it brought me to tears. I will always remember the joy and freedom I felt in making the commitment to stop drinking. It became a new memory to engulf every other one that associated alcohol with pleasure.

The past is a part of who you are today, but it doesn't have to keep you from molding your ideal future. Use this exercise to leverage your past and make it work for you.

Section 2

ACTION

"It is one thing to study war, and another to live the warrior's life."
—Steven Pressfield

Without taking action to confront our fears, all the knowledge in the world means nothing. This section will teach you the what, why, and how of Fearvana so you can use it to become a peak performer in whatever field you pursue.

Continuing with our analogy of climbing a mountain as the journey to mastery over the self, the last section taught you the art of mountaineering and to accept how little you are in control when faced with Mother Nature's fury. Now it comes time to use what you have learned to work your way toward the summit and back, one step at a time. Onward we go toward that summit.

Chapter 6

THE BIRTH OF FEARVANA

"Fear is overcome through action: sports, meditation, music, love, repetitive training, positive exposure to the object of fear, and simply using common sense to investigate what's so scary."

—Joanna Bourke

illions of stars illuminated the night. The vast expanse of the galaxy transformed my solitude into a moment of oneness with my mind, body, spirit, and the natural world around me. I paused to soak it all in. A hint of fear replaced my bliss as I glanced up at the snow-covered summits towering above me, a foreboding glimpse of the challenge that lay ahead. I silenced my mind with action and pushed onward.

Upon arriving at the base of the glacier, I took a short break to strap on my crampons. Clear breaches within the snowpack now exposed themselves, commanding fear to flow into me again. It wisely obeyed the mountain's order.

Standing tall at 17,618 feet, Pequeño Alpamayo, in the Condoriri area of Bolivia, was my first attempt at solo mountaineering. The only support I could count on had to come from within. I relished the idea of facing a new limit. It also scared me to death. I would be on the mountain for less than twenty-four hours, but that was more than enough time to send me to my grave if I lost focus.

Alone on the mountain, there would be no one to protect me from falling into a crevasse on the glacier. With each step, I faced the possibility that the snow could break from underneath me and send me plummeting into a deep, dark abyss. For that reason, most climbers prefer, if not require, walking on a glacier tied into a rope team; if one climber falls into a crevasse, the other members of the team can break the fall and pull the climber back up. One mountaineer said being alone on a glacier meant you were a "meat-popsicle"—not a pleasant thought to take with me onto the minefield.

I spent the night before the climb lying awake. Snuggled into my sleeping bag, I stared at the ceiling of my blood-red tent, lost in thought. I wondered what would become of me the next day. I questioned my decision to climb alone. I pondered ways to justify a retreat. This was a natural response to the process of testing my limits against my potential to exceed them.

Standing at the foot of the glacier was no time for the chaos of self-consciousness to enter the picture. Necessity dictated I give this endeavor all my mental energy and attention. I had to lose myself in it. I heeded the call of fear and took my first step onto the ice. Then . . . I reached Fearvana.

All of it vanished into the night—the fear, dread, doubt—everything. The awareness of my mortality dissipated with the intensity of the experience. Chaos became calm. Time became eternal. Immersed in the now, all sense of past and future disappeared. The mountain overpowered me in every way, yet I felt more in control of my fate than ever before. Here on this glistening glacier, with the countless number of unknowns ahead, nothing else mattered but the next step. The unknowns lost their influence over me. My entire world narrowed down to the few inches in front of me. I controlled what I would do within those inches.

I struggled upward in peace. The mountain and I were now in this together. For a few brief moments, as Georges Bataille said, I was like the animal who "is in the world as water in water." My human brain had disengaged. The sense of self that occurs with consciousness vanished. I was not me. I was a part of the world around me. Every step flowed effortlessly into the next one. There were other moments on the climb when the fear returned to guide my awareness back to the challenges ahead of me, such as when I walked across a narrow section with a sharp drop on both sides. That battle for my concentration was a necessary part of the overall struggle to facilitate my growth and learning. This balance between a state of pure engagement and a state of strenuous suffering is the essence of Fearvana.

To this day, I have no clear memory of that entire experience. When I look back on it, though, I know I was alive and at the peak of my abilities. I felt what it was like to experience humanity at its finest. Fear is a gift that unleashes our greatness.

Once we learn how to harness it, fear becomes an access point to the most valuable state of being in the human condition: Fearvana. Fearvana makes it worthwhile to leap into the unknown with unbridled enthusiasm. It makes the impossible possible. Fearvana is the secret to health, wealth, and happiness.

Why Fearvana?

This book combines extensive research in science and spirituality to shift the paradigm on fear, but I am just one man. It would be arrogant of me to assume I could single handedly transform our understanding of *fear*, a word so ingrained in our collective consciousness in such an unfavorable manner. Instead of attempting this losing battle of transforming *fear* from a negative to a positive connotation, I chose instead to create a new word to give us a new perspective. To give credit where it's due, my wife actually came up with the word *Fearvana*. But why bother? It's just a word, right?

Words have a lot more power than you might think. To illustrate that point, would you rather eat a piece of cooked beef that is 25 percent fat or one that is 75 percent lean? Psychologists Irwin Lewin and Gary Gaeth conducted a study in which they gave two groups of people an identical cut of cooked beef. They told one group the beef was 75 percent lean and the other that it was 25 percent fat. When asked to rate the meat before and after tasting, the group that heard the word *lean* rated the beef higher than the other group. Why?

Because the other group heard the word *fat* in the description of their meat. They did not like that word. The vocabulary used to frame the meat affected participants' perception of it.

As Dr. Andrew Newberg and Mark Robert Waldman, authors of *Words Can Change Your Brain*, state, "Just seeing a list of negative words for a few seconds will make a highly anxious or depressed person feel worse, and the more you ruminate on them, the more you can actually damage key structures that regulate your memory, feelings, and emotions."

Embracing the word *Fearvana* (or any word) and using it as part of your everyday vocabulary can literally change the structure of your brain. The word creates a new possibility for what is perceived as the greatest obstacle to success in the human experience: fear. Fearvana allows us to turn that so-called obstacle into an opportunity for constant growth.

It was necessary to create this new word to alter our relationship with fear because of a phenomenon known as the priming effect. In a study by psychologist John Bargh, two groups of students were asked to create four-word sentences using five randomly selected words. One group was given a set of words like *finds, he, it, yellow, instantly*. The other group was given a group of words like *Florida, old, gray, bingo, wrinkle*. Upon completing this task, both groups were sent to another room where they were told they would be conducting another exercise. The real goal was to measure how long they took to walk down the hall.

The group primed with words associated with the elderly took longer to walk down the hall than the other group. Both groups asserted the words they saw in the first experiment had no effect on them, yet it had a measurable impact on their actions. Their subconscious processed the words without their awareness. The same result occurred when two groups of students were primed with the words *smart* and *intelligent* versus the words *stupid* and *failure* before taking a test. Each group performed consistently with the words registered within their subconscious. Numerous other studies have validated the power of priming to control our actions. "The words that we attach to our experience become our experience, regardless of whether it's objectively accurate or not," states Tony Robbins.

Using scientific evidence, I have shown fear is very normal emotion necessary for success. Yet for some people, no matter how much I prove fear is not evil, their subconscious brain will register it as a negative emotion. It's nobody's fault. We have all been primed by hundreds of sources to believe fear is evil.

Fearvana is the clean slate. It has no past conditioning to affect it. Planting the Fearvana seed in our brain allows us to build new associations so we are no longer victimized by the effect priming has had on us regarding fear or any other "negative" emotion. The next time fear shows up, just tell yourself, "Fear is just one step away from Fearvana." Let that thought drive you forward.

The Balance between Suffering and Joy

While conducting his research on happiness, Mihaly Csikszentmihalyi interviewed thousands of people of all ages, sexes, cultures, religions, and socioeconomic backgrounds. In his findings, he discovered the means to living a full life, as opposed to floating through it until we die. He coined the term *flow* to describe this state of "optimal experience."

He defines flow as a state "in which people are so involved in an activity that nothing else seems to matter; the experience itself is so enjoyable that people will do it even at great cost, for the sheer sake of doing it." Flow is an essential element in Fearvana. As you will learn in chapter 8, the conditions that create the experience of Fearvana are synonymous with the state of flow.

However, in my research and experience, I have found success demands a real struggle to the point of questioning the very endeavor to which we commit ourselves, even if only for a moment. Joy is simply not enough to produce mastery. As I continued my years of studying what it takes to achieve greatness, I came across another concept that filled the gap left by flow and was equally responsible for the experience of Fearvana: deliberate practice.

In his book *Talent Is Overrated,* Geoff Colvin characterizes deliberate practice by the following five elements:

- It is an activity designed specifically to improve performance.
- It can be repeated a lot.
- Feedback on results is continuously available.

- It's mentally demanding, whether the activity is purely intellectual or heavily physical.
- It isn't much fun.

Flow and deliberate practice are a part of the Fearvana experience, but Fearvana was needed to bridge the gap that connects the two. Cal Newport, bestselling author of *So Good They Can't Ignore You*, describes this gap:

The knowledge work community does not yet have the right vocabulary for describing an experience that is not suffering, but then again is not immersive flow either. Work shouldn't suck. But it shouldn't feel like play either. The deliberate practice hypothesis demands that we learn to recognize (and embrace) that curious zone somewhere in between.

Fearvana is that zone.

The Gift of Fear

"I just fell in love with that feeling you get from being terrified."
—**Surfer Mark Mathews**

The development of Fearvana is a direct result of my own journey through struggle and success. I have had a very intimate relationship with both throughout my life, from my addiction to drugs and alcohol, to my time in the Marines, from building my business, to my outdoor adventures all over the world. Fear was the primal emotion at the root of these expeditions through the valleys and the summits.

When I shared some of these stories during my first public speaking engagement at the Verizon headquarters, someone asked me how I became fearless. I told her I'm not. I've felt fear every time I did anything worthwhile in my life. She looked surprised that fear was so much a part of me.

Young students I speak to often have this false sense of bravado, saying they are not afraid of anything. To most people I share my stories with, they seem to think the presence of fear means cowardice. A friend I once worked with told me, "I just need to wait for the fear to go away so I can quit my job." His problem was not the fear; it is very natural to be scared to quit a job to start a business. His problem was that he made the fear mean he was weak. Like most of us, he believed fear was a negative thing. I myself forget this from time to time.

When fear showed up once while I was rock climbing in New Jersey, I remember saying things to myself like, "What is wrong with you? What are you afraid of? Shut up and stop being a baby!" My first climb that day was a disaster.

Before the second climb, a realization hit me: "Wait a second; I am writing a book about this!" In an instant, everything changed. I stopped fighting the fear. I became present to my anxiety, nervousness, ego—everything. I breathed it all in and became one with it. Every inch up that wall played out like a graceful ballet. Fear is never the problem. It's the fear of fear that cripples us. As Sir Richard Branson said, "It is important not to fear fear, but to harness it—use it as fuel to take your business to the next level. After all, fear is energy."

Harvard professor Alison Brooks conducted an experiment with two groups of people preparing for a speech. She found the speakers who relished the fear by telling themselves, "I am excited," not only felt more confident, but they also performed better than the speakers who tried to relax and tell themselves, "I am calm." Both groups of speakers felt nervous before their talk, but one group harnessed that energy, while the other tried to eliminate it.

The very act of fully embracing fear creates the necessary level of composure to leverage its power. Being afraid of fear or running away from it is precisely what sends us deeper and deeper into a pit of despair, which becomes harder to climb out of the more we dig ourselves into it. When we fight or resist fear, we throw away precious mental energy in the downward spiral of second dart syndrome.

To be clear, I am not suggesting there is no value in the state of calm. When you are receiving insight about techniques and strategies to perform better, it helps not to have the amygdala activated by fear so that you can use your human brain to process the information. Even in Marine Corps boot camp, skills and techniques are taught in a surprisingly tranquil setting.

I would imagine you are reading this book in a safe and comfortable environment. This kind of setting allows for learning to take place. It would be near impossible to soak in all of this information if someone was shooting at you.

But when it comes to implementing these principles, that's when the fear will strike. That's when you want it to. When it comes time to perform and you feel those butterflies in your stomach, don't push them away or try to suppress them. Embrace them. By channeling the positive energy of fear, we can use it toward a more productive means.

A graduate student of Dr. Sian Beilock, a professor of psychology at the University of Chicago, conducted a study to determine how stress would affect students' response to a math test. After the test, he found some students performed well, despite high levels of cortisol (the stress hormone), while other students performed poorly in response to equally high levels of stress. The difference was that the high-performing group did not define themselves as people who got anxious over math. On the other hand, the group that performed poorly reported that math made them extremely anxious. In her response to the study, Beilock said, "Interpreting the situation and your bodily response in a positive rather than a negative light may be a key to performing well when it counts the most."

The obstacle to peak performance was not heightened stress levels; it was the students' belief about themselves in response to the stress and what those stressful feelings meant. High performers don't stress out when they feel stress.

You know now that fear in any form, whether it is heightened stress or anxiety, is a natural response to external stimuli beyond our conscious control. There is nothing you can do about its presence, at least not initially and without awareness. It is our perception and understanding of fear that affects how we respond to it.

In a study by social psychologist Dr. Jeremy Jamieson, it was found that embracing fear, instead of trying to eliminate it, works even for people who suffer from severe social anxiety disorder. After teaching the participants in his study the value of our body's automatic stress response, he told them, "When you feel anxious or stressed, think about how your stress response can actually be helpful." He then placed them in a mock interview designed to induce high levels of stress, especially for people who struggle in social situations. It's called the "social stress test."

Learning how to harness their brain's natural stress response didn't reduce the amount of anxiety participants experienced, but it allowed them to choose a new meaning for that anxiety. In fact, the participants who measured higher stress levels reported a higher level of confidence and performed better in the test, as rated by third-party observers. That is the power of Fearvana. It gives you a way to reframe your attitude toward your fear so you can make it mean something empowering for you the next time it shows up.

Take the example of UFC star and former champion Ronda Rousey. In her thirteen professional UFC fights, she holds the record for the fastest victory in UFC history: a mind-blowing fourteen seconds. That is just one among her other notable achievements, such as finishing opponents in an average of two minutes and sixteen seconds and being the only UFC fighter to win three title fights in less than a minute. This is how she describes her relationship to fear:

> I'm full of fear, without any doubt. I'm always f****** scared. You have to have fear in order to have courage . . . My fear of failure is larger than anyone else's, and it's increasing with every single fight. I don't shy away from it. I am extremely fearful, but I don't have a single doubt in my mind [when it comes to winning] . . . That's the environment I do the best in. I fight above myself. I do better when the pressure and the fear is the highest.

When Rousey lost her first professional fight and the title to Holly Holm, Holm told the world, "I fear everything about her." Fear made these two women warriors.

Neurologically, we are wired to seek out the fear that the quest for Fearvana demands. This makes sense evolutionarily speaking. Our ancient ancestors faced a far more unpredictable, albeit simpler, world than we do today, so the brain responded in a manner that helped them navigate all the unknowns of their environment, rewarding them every time they did, thus ensuring their survival.

Remember what we discussed in the introduction? Excessive security and comfort is a burden. In a dangerous world that warranted life-threatening fear, Fearvana was a given. We don't live in that world anymore, so we must seek out fear to realize our greatness, achieve our goals, and live a higher quality of life. For me, writing this book was the ultimate Fearvana experience. It was beautiful, stressful, and terrifying. The frustration of deliberate practice and the pleasure of flow merged in Fearvana.

Through the experience of Fearvana, I managed to write, and finish, this book about it. The struggle was absolutely worth it. I no longer want to see so many people victimized by fear, stress, and anxiety simply because of a monumental misunderstanding of what they are. These gifts do not have to be reserved for only an elite few.

Chapter 7

THE MINDSET OF FEARVANA

"Only those who will risk going too far can possibly find out how far one can go."

—T.S. Eliot

I magine millions of people with their eyes plastered on you as you set foot on a basketball court for the final game. Your team depends on you. The entire city is counting on you. Lose this game and history will forever remember you as the sole reason for banishing your team to the realms of obscurity. If you found yourself in this position, what would you do?

A) I would never have put myself in this position in the first place. It sounds terrifying!
B) I would vomit.
C) I would try to calm my nerves.

This is how Bill Russell responded to that scenario. Bill Russell is the only person to win a college basketball championship, an Olympic gold medal, and an NBA championship—all in the same year. During his NBA career, he led the Boston Celtics to eleven championship victories in thirteen years. He is often regarded as one of the greatest players in the history of the sport.

His greatness came at the cost of a very uncomfortable pre-game ritual. While waiting in the locker room, Russell felt so overwhelmed with anxiety that he would run to the toilet to throw up. The Celtics coaching staff called in a doctor to check if Russell suffered from dehydration, but nothing appeared to be wrong.

One day, near the end of the 1963-64 season, Russell entered the locker room and felt relaxed. For the first time in his career, he did not vomit before a game. It led to the worst performance of his life. As the season progressed, he continued to endure the burden of calm before every game. The media wrote headlines such as "Russell's Slump Causes Celtics Another Loss," and "The Legend Finally Loses His Touch."

Through consistent effort from his teammates, the Celtics managed to make it to the playoffs anyway. Beaten down by his recent slump, Russell arrived early for Game 1 of the series of seven to avoid the fans and the media. Fortunately for Russell, they got there before him. Watching the hordes of Celtics fans converge around the stadium flooded his nerves. The cumulative pressure reminded him of his early days in basketball as a rookie, fighting an uphill battle. He entered the locker room and bolted for the toilet. Fear found him again. "We're going to win, guys! We're going to win," he told his teammates. That season the Celtics took their eighth title.

The Superhuman Chemical Cocktail of Fearvana

"Viewing the stress response as a resource can transform the physiology of fear into the biology of courage."

—**Dr. Kelly McGonigal**

Rest assured, you don't have to be scared to the point of throwing up to turn fear into your ally. You simply have to accept fear is not your enemy. In fact, our brains release chemicals during Fearvana that indicate we are designed not just to handle scary situations, but to thrive in them. There are six chemicals released during Fearvana.

Dopamine

Dopamine is released any time we do something new, making it a particularly strong candidate to demonstrate we are wired for Fearvana. Why else would our brain secrete the "feel-good" hormone when we take a risk? Novelty makes us happy. We need variety to keep us engaged with life.

In her research on neuroscience, Sharon Begley found the "inability to see things as new, as fresh, as different, this is what elicits the feeling of depression." I have found this to be true with all my clients who have suffered from moments of depression. When they started taking risks, however small, they opened their world to new experiences, which radically improved the quality of their lives.

Pushing into fear leads to existential and neurological happiness. It's as if our brain is telling us to become explorers of the unknown. The release of dopamine during Fearvana also helps us focus our attention on navigating that new world.

Adrenaline

Adrenaline is one of the first chemicals released by your brain at the outset of fear and danger. Too much adrenaline leads to panic, but if you get just the right amount, it starts to work for you. During Fearvana, adrenaline helps to increase your energy, improve your strength, enhance your focus, sharpen your reflexes, and boost the efficiency with which your body and mind operate.

If you are attacked by someone in a dark alley, adrenaline will help you shut out external distractions to focus only on your attacker. The more you are trained in Fearvana, the greater emotional control you will have to leverage the adrenaline so you don't freak out when you see the knife. It will then stop blood from flowing into your digestive system and increase the flow into your heart and muscles where it is needed. This will make you fight harder. The release of glucose triggered by the adrenaline will give you more energy to fight longer as well. Adrenaline turns us mere mortals into masters at navigating unknown territory.

Endorphins

Ever heard the term "runner's high"? What that refers to is the release of endorphins, which are partly responsible for the state of bliss experienced during Fearvana. Endorphins help us shut out pain during those moments when we don't need it distracting us from acting beyond our limits. They are so good at blocking out pain that one particular kind of endorphin is known to be even stronger than morphine.

Anandamide

Endorphins contribute to the bliss in Fearvana, but anandamide is the primary player. The name itself comes from the Sanskrit word *ananda*, which translates to bliss. Anandamide was discovered in 1992, so there is still a great deal of work being done to better understand its functions, but according to Steven Kotler, it "augments lateral thinking, producing connections between much farther-flung ideas." Anandamide gives you an enhanced ability to think your way out of dangerous situations.

Furthermore, according to scientists at Duke University, anandamide could help "facilitate the extinction of traumatic fear memories." Anandamide doesn't just help you handle fear when it shows up; it also helps eliminate the source of that fear in the future.

Serotonin

Serotonin is the neurotransmitter best known for its role in depression. Scientists are still unclear whether lower levels of serotonin lead to depression or whether depression causes a reduction in serotonin. However, it is clear serotonin has a strong influence over our moods and emotions, and low levels of it are not a good thing. Professor Philip Cowen from Oxford University says, "It's a molecule involved in helping people cope with adversity, to not lose it, to keep going and try to sort everything out."

Serotonin also contributes to feelings of significance and importance, which helps increase our confidence to face the challenges inherent to Fearvana.

Stress

When we find ourselves in a stressful situation, such as the moment we step outside our comfort zone, two hormones are released to help us combat the stress: cortisol and dehydroepiandrosterone, or DHEA.

Cortisol is the "stress hormone." This chemical has a bad reputation, but it isn't all bad. When harnessed, cortisol works in a manner very similar to adrenaline. It helps your body temporarily ignore the functions it doesn't need so it can channel all its energy to the task at hand. As Kelly McGonigal writes in her book *The Upside of Stress*, "When it comes to performing under pressure, being stressed is better than being relaxed." However, also like adrenaline, extremely high levels of stress for a long period of time can be damaging. This is evident in people with severe post-traumatic stress disorder.

DHEA is cortisol's counterpart. Evolution blessed us with this neurosteroid to help our brains learn from stress and become tougher as a result of it. This doesn't mean DHEA is always the good guy and cortisol is only sometimes our friend. We need the right ratio of the two hormones to most effectively deal with a high-stress environment. That ratio is known as the growth index. The higher our growth index, the better we become at responding to stress.

"The DHEA-S-cortisol ratio may index [indicate] the degree to which an individual is buffered against the negative effects of stress," said Dr. Charles Morgan from the National Center for Post-Traumatic Stress Disorder. So how do we control that ratio to ensure it works in our favor? We embrace the mindset for Fearvana. These six chemicals that make up the rocket fuel released by fear will only work for us if we believe in their ability to do so.

Our Most Powerful Weapon

"Whether you think you can or you think you can't, you're right."
—**Henry Ford**

In one 1980 study, researchers used makeup to apply a ghastly scar on the faces of the participants. After looking at their faces in the mirror, they were instructed to engage in a conversation with another person and notice whether the scar made an impact on how they were treated. Before leaving the room, the researchers said they needed to touch up the scar to ensure it did not fade. In reality, they removed the scar and no longer allowed the participants to look in a mirror.

After their conversations, the participants still felt they were "treated horribly and that the other person was tense and patronizing," writes Professor Gazzaniga in his description of the experiment. They interpreted common human interactions, such as a brief glance in another direction, as rude because they believed their face to be disfigured. Confirmation bias ensured any incoming information was in line with that belief. Think about all those inner scars we have that shape our perception of how others relate to us and, more importantly how we relate to ourselves.

Some time ago, I used to berate myself after every run. If I ran five miles, I got angry I hadn't run seven. If I ran ten miles, I got upset for not running twelve. No matter what I did, it was never enough.

Ever since coming home from Iraq, I have struggled with the belief that I did not do enough in the war, that I should have died out there. This belief about my self-worth painted a lens of not enough that affected almost all other areas of my life.

By practicing the self-awareness exercises I have shared with you, I brought this limiting belief from the implicit to the explicit. From there, I consciously worked on creating a new, empowering belief. Without self-awareness, we can all become victimized by the negative beliefs embedded into our animal brains by our environment.

Harvard professor Robert Rosenthal conducted a study in which he told a group of elementary school teachers that a small section of randomly selected students were academically gifted and about to experience significant intellectual growth. The teachers' beliefs led them to behave differently with the students they thought were gifted. They engaged with them more; they touched and smiled at them more; they gave them more time to ask questions; they gave them better feedback; and they provided them with more acknowledgement.

Their behavior had a direct impact on their students' performance. "If teachers had been led to expect greater gains in IQ, then increasingly, those kids gained more IQ," Rosenthal said.

The teachers did not consciously treat their students differently. Their subconscious beliefs triggered subtle actions that caused a significant ripple effect in the lives of their students. Of course, the kids did not have the ability to separate themselves from their environment, so they had no control over how their animal brain responded to the teachers. Their external world shaped their destiny, as it does for so many adults too.

Our beliefs are molded by a variety of forces, from past experiences, to television, to books, to advertising, to other people, to the Internet, to pretty much any external stimuli we encounter. To choose our own beliefs, we must consciously decide how we interpret that stimuli.

Nelson Mandela spent twenty-seven years in prison and still forgave the very people who put him there. He chose to find an empowering meaning for the experience, which led to the belief he must work together with his former captors to create a better home for all South Africans. His beliefs changed the face of the nation. Such is the power of belief.

Beliefs are the convictions we hold to the meanings we assign to everything around us. They are a feeling of certainty in the correctness of those meanings. According to Dr. Michael McGuire, a professor at UCLA, a belief is "a state or a habit of mind in which trust or confidence is placed in some person or thing. It is to accept as true, genuine, or real. It is to have a firm conviction as to the goodness, efficacy, or ability of something."

Beliefs set the framework from which we view ourselves, others, and the world around us. If you wore glasses with a red lens, your world would look red. Beliefs are that lens. They are responsible for creating our entire sense of reality. This book is all about the power of the mind to shape our world, internally and externally. Beliefs are the strongest weapons we have to harness that power.

While treating wounded soldiers during World War II, Harvard doctor Henry Beecher ran out of anesthesia due to massive casualties. Without any other available options, he began administering salt-water injections to treat his patients, but they believed they had been given anesthesia. This belief gave them the superhuman ability to endure the excruciating pain of invasive surgery without flinching.

Similarly, neuroscientist Helen Mayberg discovered that activity in the brains of patients with depression changed in the same way, whether they took an antidepressant medication or a placebo. Their beliefs influenced their brain on a neurological level.

One particularly fascinating study involved one hundred students who were divided into two groups. Each group took one of two different drugs to test their effectiveness. One group was told they were taking a stimulant. The other was told they were taking a tranquilizer. In reality, each group was given the opposite. Remarkably, half the students displayed symptoms in line with what they had been told, which was in direct contrast to the chemical substance flowing through their body. In response to the study, Dr. Beecher said the value of a drug "is a direct result of not only the chemical properties of the drug, but also the patient's belief in the usefulness and effectiveness of the drug." This phenomenon is known as the placebo effect.

A firmly held belief can overpower almost all other forces when it comes to molding our reality. People fight, die, and even kill for such beliefs. The content and the source of these beliefs are irrelevant, as demonstrated by the placebo effect. To affect any kind of change, all that matters is that a belief has taken root. "There is an innate capacity for our bodies to bring into being, to the best of their ability, the optimistic scenarios in which we fervently believe," states Harvard professor Anne Harrington.

When the legendary mountaineer Reinhold Messner set out to climb Mount Everest without oxygen in 1978, doctors told him it was suicide. They said no human being could survive at that altitude without supplementary oxygen. When he returned from the summit, he was asked why he went up there to die; he said, "I didn't. I went up there to live."

A study of athletes competing in the Olympics or the Commonwealth Games found the most common definition of mental toughness was "having an unshakable belief in your ability to achieve competition goals." Your beliefs can triumph over any internal or external conditions that stand in the way of your success. My belief in my own ability gives me the strength to run ultramarathons, despite my own biological defects. Not only do I have a blood disorder called thalassemia, I also have mild scoliosis, flat feet, and weak knees. But biology is not destiny. Belief is.

"To perform your maximum you have to teach yourself to believe with an intensity that goes way beyond logical justification," says one of the greatest soccer managers of all time, Arsene Wenger. "No top performer has lacked this capacity for irrational optimism; no sportsman has played to his potential without the ability to remove doubt from his mind."

How do you create that belief in yourself and your ability if you don't currently possess it? Beliefs are as real in your brain as anything else, so they are also as malleable. No matter what your current beliefs about yourself or the world around you, the three principles of neuroplasticity ensure you can alter those beliefs to bring them into alignment with the person you want to become. The road to creating that person starts with one essential belief.

The Most Important Belief of All

"Beliefs are the key to happiness [and to misery]."

—**Dr. Carol Dweck**

The first step to reshaping all other beliefs is to ingrain a fundamental belief regarding the nature of human potential: talent, skill, and ability are not inborn; they are forged through training. This one belief is the umbrella belief upon which every other one depends. It is the magic of the physical brain's neuroplasticity translated into the thinking mind's correlate. If you don't believe you can change yourself with hard work, you won't take action, and no other belief will matter.

Psychologist Dr. Carol Dweck leads a great deal of the research conducted on the driving forces of motivation and human potential. Through her work, she discovered the two mindsets (beliefs) that separate successful people from those who never achieve their goals.

Successful people prescribe to the growth mindset, which according to Dr. Dweck, is "based on the belief that your basic qualities are things you can cultivate through your efforts. Although people may differ in every which way—

in their initial talents and aptitudes, interests, or temperaments—everyone can change and grow through application and experience." A growth mindset is one that believes every experience is an opportunity for growth and human potential is limited only by our imagination.

On the other hand, unsuccessful people abide by the fixed mindset. When a person with a fixed mindset encounters an obstacle, instead of learning and growing from it, he either blames some outside force, and/or he allows it to lower his self-esteem. Someone with a fixed mindset believes one's qualities are innate and unchangeable. A major obstacle to Fearvana is the widespread acceptance of the fixed mindset.

Roger Federer was said to have tennis encoded in his DNA. People have said Tiger Woods was born to play golf. At a young age, we are programmed to believe we are born with certain innate abilities that establish the limits of our potential. I still remember being excluded from the gifted and talented program in high school because it was only reserved for a select few students who possessed a "gift" or "talent" other students, like myself, did not. Unfortunately, programs like this can be found in schools across the globe.

This is not to say genetic differences do not exist; of course they do. Tiger Woods learned to balance on his father's palm when he was just six months old. Most of us can barely stand at that age. Ed Viesturs, the first American to climb all fourteen mountains over eight thousand meters, has a lung capacity of seven liters—two more than the male average. This makes it a lot easier for him to breathe in high altitude.

Along with physical, genetic differences, there are mental ones as well. Some people are born with naturally lower levels of dopamine, making them more prone to addiction. But no matter what hand has been dealt to us, we get to choose whether to complain about it or do something about it. Our predetermined fate is never the real problem. The problem is when we respond to Lady Luck with the belief that our genes make us who we are.

Even if those genes are better than average, they mean nothing unless we take action to consciously shape our own destiny. We can't control fate, but we can control how we choose to respond to it. Tiger Woods and Ed Viesturs would not have achieved any success without putting in hours of hard, focused training at the edge of their abilities. This rule applies to anyone, anywhere—regardless of genetic conditioning.

In 1991, the father of deliberate practice, Anders Ericsson, and his team went to the Music Academy of West Berlin to find out what separates peak performers from everyone else. He split the violinists in the school into three distinct groups, based on interviews with professors and the success each student experienced in competitions outside the school.

The first group consisted of the top performers. They were expected to become world famous solo violinists. The next group consisted of students just below the abilities of the first group. These students would probably end up playing in top orchestras, just not as soloists. Finally, the third group was comprised of students in a course with significantly lower standards than the other groups. These students were clearly not as skilled as the others.

Through extensive interviews, Ericsson found the students in all three groups shared very similar backgrounds. Each began formal training when she was eight years old. She made the conscious decision to become a musician before the age of fifteen. Each was trained by an average of 4.1 music teachers. The three groups appeared to be identical on almost all accounts except one: training.

By the age of twenty, the top performers had spent an average of ten thousand hours training, two thousand more than the second group and six thousand more than the third. He discovered that not a single top performer reached that level without the ten thousand hours of training. Not one! He first discovered the ten-thousand-hour rule Malcolm Gladwell is now known for. "The differences between expert performers and normal adults are not immutable, that is, due to genetically prescribed talent," Ericsson says. "Instead, these differences reflect a life-long period of deliberate effort to improve performance."

It is said that Shizuka Arakawa from Japan, often considered one of the greatest skaters in history, fell more than twenty thousand times in her nineteen-year journey from amateur to Olympic gold medalist. She persevered because she believed every fall would lead to growth. Nadia Comaneci was the first gymnast to win a perfect score of ten at the young age of fourteen. Her road to greatness began in kindergarten when she first started training. Through years of rigorous hard work, she too earned an Olympic gold medal.

There are countless examples of people who defied their own limitations through intense training. As a child, Albert Einstein was deemed "sub-normal." He did not speak until the age of four and could not read until he was seven. He was told by his teachers he was "mentally slow, unsociable, and adrift forever in foolish dreams." Consequently, he was expelled from school.

Thomas Edison's teachers told him he was "too stupid to learn anything." Ludwig Van Beethoven's teachers called him "hopeless as a composer." As a sophomore, Michael Jordan was cut from his high school varsity basketball team. Charles Darwin was ridiculed for being lazy. He wrote of his early years, "I was considered by all my masters and my father a very ordinary boy, rather below the common standard of intellect." Lucille Ball, the award-winning actress who became famous for her role in the hit show *I Love Lucy*, was told by her instructors in drama school to "try any other profession." Early in his career, Elvis Presley was fired after just one performance. His manager told him, "You ain't going nowhere, son. You ought to go back to driving a truck." Oprah Winfrey was fired from an early job because she was "unfit for TV." Walt Disney was fired because he "lacked imagination and had no good ideas." The list is endless.

Every one of these extremely successful individuals worked incredibly hard to improve their skills because they believed in the growth mindset and acted on it. The more we value perseverance and hard work, the more likely we are to achieve mastery. Life will always hand us adversity. When we believe in the value of effort over innate ability, we gain the confidence to face that adversity head on and respond to it as a challenge to be overcome.

The impact of cultivating this belief is monumental, especially at a young age. In one study, Dweck and her team gave 330 fifth and sixth graders a questionnaire that determined whether they believed in the fixed mindset or growth mindset. The students were then given a set of twelve problems to solve. The first eight were easy and well within their abilities; the last four were significantly more challenging. Dweck provided ample motivation for the students to solve these problems by offering each one a reward of his or her choice.

As soon as the problems became too difficult to solve, the students with a fixed mindset began saying things like, "I guess I'm not very smart," "I never did have a good memory," and "I'm no good at things like this." Despite finding success in the previous eight problems, they immediately lost faith in their own abilities and allowed an obstacle to affect their sense of self-identity.

While the fixed-mindset students blamed their own intelligence when they encountered failure, the growth-mindset students "did not blame anything." Dweck recalls, "They didn't focus on reasons for the failures. In fact, they didn't even seem to consider themselves to be failing."

Along with their attitude, two thirds of the fixed mindset students displayed a clear decline in problem solving strategies once the questions got too difficult. On the other hand, more than 80 percent of the growth mindset students either maintained or improved their strategies. A quarter of the group even taught themselves more advanced strategies.

The gap in performance between the two groups was not a result of intelligence, skill, or motivation. It was a belief in their own ability. When the growth-mindset group met with failure, it simply meant they needed to work harder. They chose to interpret struggle in an empowering manner that did not negatively affect their sense of self-identity. The group that valued effort over inborn talent believed what Thomas Edison already knew: "Genius is one percent inspiration and ninety-nine percent perspiration."

Dweck even demonstrated the impact of implanting such a belief in young students through the transformational power of language. After assigning a group of 400 fifth grade students a series of simple problems to solve, Dweck gave them their score and then praised them for it. Half the students were told, "You must be smart at this." The other half was told, "You must have worked really hard." Those six simple words radically affected future performance.

When asked whether they would rather a take a tough or easy test following the first one, 90 percent of the students who were praised for effort chose the harder test, while two-thirds of the students praised for intelligence preferred the easier test. The group praised for their talent wanted to maintain their intelligent status while the group praised for effort wanted to test themselves through a greater challenge.

When given problems they could not solve, the effort-praised group worked harder and longer before quitting. More importantly, they did not let the failure affect their confidence and sense of self. Finally, when the students were given another test equal in difficulty to the first one, the students who believed in their intelligence showed a drop in 20 percent while the group that believed in effort improved by 30 percent. Three additional experiments by Dweck validated these findings. "Praising children's intelligence harms their motivation, and it harms their performance," said Dweck in response to these studies.

Whether you want to be a climber, writer, entrepreneur, painter, or anything else, being a peak performer in any field starts with a belief in the growth mindset. But implanting this belief is only possible if you first accept beliefs are malleable. If you choose to be rigid in your view of the world, even if that perspective isn't serving you, none of the training exercises below will work. However, if you are willing to be open to new possibilities for yourself, the transformation available to you is magnificent.

"Small shifts in mindset can trigger a cascade of changes so profound that they test the limits of what seems possible," writes McGonigal in *The Upside of Stress*. "Changing our minds can be a catalyst for all the other changes we want to make in our lives." This entire book is one big mindset shift exercise. The series of studies I share and the stories I tell are meant to plant seeds in your subconscious that will permanently shift your relationship with fear, stress, and anxiety.

In a study by psychologist Alia Crum, two groups of participants viewed a video about stress before being sent into an interview designed to elicit a high stress response. One group saw a video about the positive effects of stress while the other saw one that emphasized its negative effects.

During the interview, cortisol levels increased in both groups of participants. But the participants who were exposed to the idea that stress was enhancing displayed a measurable increase in DHEA and a higher growth index than the participants who were taught stress is damaging. Their belief about stress governed their reality in an objective and quantifiable way.

If a single, three-minute video about the positive effects of stress can cause a physiological shift inside of us, imagine what an entire book can do for you. The right belief will create a snowball effect in all areas of your life that will stay with you forever. By drilling into your animal brain and implanting these new beliefs deep into your subconscious, we are tapping into the source of your thoughts and actions. The new perspective will then circulate throughout all parts of your consciousness, creating a new you. When the right seed is planted, it will grow out its roots and blossom. Let us plant that seed.

Training Exercises

Here are two separate exercises to help you create new beliefs and mindsets. One is no better or worse than the other; they're just different. Try both if you like, and see which one works best for you. My intention is to give you as many weapons as possible to use on your journey to becoming the person you want to be.

Training Exercise 1: Molding Your Mindset

1. Learn a new mindset (such as "fear is positive") by finding or exposing yourself to a different perspective, as you have done by picking up this book. In one study to help incoming freshmen students at an Ivy League school feel like they belong, psychologist Greg Walton first exposed the students to surveys from juniors and seniors who also struggled with social belonging. By doing this, the new students suddenly realized they were not alone in feeling this way.

2. Do something to embrace the new mindset. You could write down a time in your life when you found yourself aligned with that mindset or take on any action that allows the mindset to grow its roots into your subconscious. To continue from Walton's example, he then asked the students to write down how their challenges were much like the ones the juniors and seniors had faced.

3. Share this new mindset with others and teach them how to better their lives with it. Giving this knowledge away to serve other people helps ingrain it into your own subconscious. For example, Walton asked the students to help next year's freshmen by sharing their stories in a video.

4. Keep taking actions that allow you to consciously practice the new mindset. With regards to fear, put yourself in stressful or scary situations and use the LMNOP cycle to continue finding the value in such emotions.

Walton's mindset intervention consisted of the first three steps only. I added the fourth as an extra measure of reinforcement to establish a new mindset. But for the freshmen in Walton's study, even those three steps led to improved health, happiness, and academic performance throughout their college life.

Training Exercise 2: Calibrating Your Compass

1. Use your growing awareness to identify your limiting beliefs. Write down as many of them as you can think of.
2. Choose one to work on.
3. Recognize and accept this belief is not a fact. This belief is something your brain has created subconsciously without your awareness.
4. Identify and write down the subconscious and conscious forces that have you holding on to that belief. For example, I held on to a belief of not being worthy enough because I felt like it should have been me that died in Iraq instead of my friend.
5. Remember there are two driving forces of all human behavior: pain and pleasure. Identify the pain and pleasure of letting go of this belief. The following are questions to help you find the source of pain:

   - What is predictable if this behavior continues?
   - What is holding onto this belief costing me?
   - What are the consequences?
   - What is it preventing me from getting?
   - What will it cost me if I don't change?
   - What is it already costing me?

Next, use these questions to find the source of pleasure:

   - What is possible in my life if I do change?
   - What could I accomplish if I shifted this belief today?
   - How much happier could I be?
   - How will creating a new belief make me feel about myself?
   - How will this change enhance all areas of my life?

6. Find the evidence that led to this limiting belief. Look back into your past and figure out what event led to the creation of that belief. You weren't born with it; it came from somewhere. Using some of the exercises from earlier in this book, travel back in time to find the source of your belief.

7. Reframe the evidence. Once you find the event that led to the belief, ask yourself what meaning your subconscious created for it. Then choose a new meaning for that event, just as Rachel did when she realized her friends gave her "the wings to fly."

8. Gather new evidence to fit the new belief. Find references in your own life, in books, from people around you—everywhere —to validate your new belief. If I wanted to climb Mount Everest without oxygen, I could read stories about people who did that to prove to myself it is possible. Of course, this would only work for me if I first believed in the growth mindset and the idea that if one person can do something, anyone can.

9. Condition the new belief by taking action in service of that belief and continuing to find evidence to validate it.

Chapter 8

THE EXPERIENCE OF FEARVANA

"In the midst of movement and chaos, keep stillness inside of you."
—Deepak Chopra

T hrough my research on fear and the manner in which successful people push through it, I discovered ten factors that create the complete Fearvana experience. Similar to how our two brains communicate, there is some degree of fluidity in this. Don't get too fixated on the rigidity of the sequence outlined below. What matters is that you are aware of these ten factors so you can take the actions that lead to your growth. These factors occur in the following three phases:

- Pre-action: This phase describes the process required to set the stage for the Fearvana experience.
- Action: This phase will outline the conditions that occur during the actual state of Fearvana, illustrating we are not just neurologically wired for Fearvana, but psychologically and spiritually primed as well.

- Post-action: This phase ensures the Fearvana experience leads to growth of the self.

Pre-Action

1. Conscious Assessment of the Risk and Fear before the Action

To experience Fearvana, the human brain is paradoxically required to turn itself off. In other words, it takes consciousness to destroy consciousness. Big-wave surfer Laird Hamilton states the necessity of awareness in taking risks: "People ask me if I feel fear in the big waves. Of course, I'm afraid. If I'm out in fifty-foot surf and not feeling fear, then I'm not properly assessing the situation." Without conscious assessment, Hamilton's decision to wrestle the thundering ocean would be an act of insanity.

Similarly, it has been said of warfare that anyone who says he is not afraid is either lying or insane. "You don't want someone without a fear response at all," said neurobiologist Dr. Lilianne Mujica-Parodi about her work with war veterans. "That's not brave; that's just abnormal." According to Mujica-Parodi, the ideal response to fear is one where the brain consciously assesses the risk, deals with it effectively, and is then able to rapidly return to a state of equilibrium once the threat is over. She calls this a warrior brain.

"Rationally addressing the cause of fear can allow you to go for it instead of maintaining strict control. And in that going for it you unlearn self-limiting patterns," says Mark Twight, the legendary alpinist and founder of Gym Jones, which is called "The World's Best Gym" by *Men's Health*.

"Extreme" athletes like Twight are not foolhardy. They are practitioners of the warrior brain. Even when the risk of death is high, they do not seek death; they court it to come alive. They make conscious choices to experience the highest form of bliss and unrestricted access to Fearvana. Such access requires the conscious assessment of risk because that allows them to determine the context of the fear, explore its potential for growth, and choose an action accordingly. Without a conscious assessment, we become victimized by fear and the animal brain beyond our control. I could have stayed away from Barbados after reading about a robbery there. Instead, I looked beyond that one incident to analyze everything that trip would entail.

To engage your fears, author Tim Ferris suggests asking yourself, "What's the worst that could happen?" This question allows you to rationally process the fear. In many cases, you will find the answer is never as bad as your emotionally charged animal brain would have you believe. Only by stepping outside of the fear to understand it can we then acknowledge its presence and do something about it. As Sir Richard Branson said, "Entrepreneurship requires a special kind of courage—you must face a great deal of uncertainty as you launch and maintain your business. The ability to recognize your fears, assess the causes, and then make decisions about how to proceed can mean the difference between success and failure for a new company."

2. Clearly Defined, Consciously Generated Goals

At the time of this writing, there are numerous debates in the personal development community about whether goal setting is bad or good. Ultimately it doesn't matter because we are always setting goals anyway. We set most of them subconsciously, but we still set them. In reading this book, you are setting a goal to finish this sentence, then the next, and so forth. By walking down the street, you set a goal to get to the next lamppost. When you watch television, you are subconsciously setting a goal to finish a show or movie. In every moment, our brain is setting a goal to get us to the next one.

"Our brains are always looking for a cue as to where to spend energy now," says Yale psychologist John Bargh. "We're swimming in an ocean of cues, constantly responding to them, but like fish in water, we just don't see it." Instead of having our goals chosen for us by our animal brain, we are now going to take control and choose them from a place of awareness. Consciously generated goals give purpose to action.

The author's purpose is to finish writing a book. The mountaineer's is to reach the summit and return alive. The athlete's is to win. To achieve the larger goal, we set smaller, specific targets like finishing a chapter, reaching the next ridgeline, putting the basketball in the hoop, or taking the next step. Psychologist and author of *Overachievement,* Dr. John Eliot calls this target shooting.

Marathoners don't think about mile twenty-six when they start; they think only about the next step. This narrows focus, preventing the human brain from intervening, with all its doubts and worries, about how miserable the rest of the run will be. When you zone in on a target, that target becomes your life. Every time you reach one, you have a new one to devote the entirety of your being to.

Engaging our human brain to set clear directions makes it easier for our animal brain to follow those directions. Setting a target to aim for gives the bodyguard clarity on what external stimuli to allow into the brain to accomplish the desired action without the conscious mind. This is why an athlete can ignore the noise of a stadium to win the game.

Describing his drop down a one hundred-foot waterfall, whitewater kayaker Ed Lucero said, "I just focus on where I'm going. I don't see the people along the banks anymore. I'm just thinking about my line. Everything is really simplified." When you direct your mind with clarity and focus to a specific action, over time, that rewires your brain, making the behavior automatic. Target shooting gives rise to mastery by allowing the quantum Zeno effect to do its job.

Clear goals also provide continuous feedback as to whether or not we are achieving those goals. In creative pursuits, such as writing this book, the feedback is a little vaguer than in the more active Fearvana experience of climbing a mountain. In more passive kinds of activities, "a person must develop a strong sense of what she intends to do," writes Csikszentmihalyi.

In writing this book, the feedback came in the form of asking myself something like, does this paragraph convey my message in the way I want it to? In the outdoors, the feedback is much easier to assess. While climbing, I am constantly getting feedback from the snowpack, the mountain, the weather, my stamina, my mindset—everything. This stream of feedback from the environment allows me to adjust my actions as I go about achieving my goal.

A clearly defined goal also creates the space for order within the chaos of consciousness. Without such clarity, our lazy brains retreat to the easiest course of action, which generally means doing nothing. This is why most people struggle to achieve their goals. I cannot tell you how many clients I have worked with who came to me with vague goals, such as "I want to lose weight" or "I want to make more money," but they have no idea how much weight or how much money they are aiming for.

After interviewing thousands of entrepreneurs, Jack Canfield found clarity is one of three traits shared by all the successful men and women. The other two traits were taking 100 percent responsibility and possessing the ability to constantly be in action. Factor 1, the assessment of risk and fear, helps us set a goal, but a clearly defined, consciously generated goal also allows us to assess the risk and fear. These two factors do not need to take place in this order, but it is important they both take place so we can then take action.

3. An Action Prefaced by Fear and the Element of Risk

The action you choose to take doesn't need to generate an overwhelming or clear-cut fear experience. Remember, fear, stress, and anxiety are all part of the same neurological process, so any one of the three can preface the action.

There is no universal unknown. It can be anything for anyone. If the idea of risking your life to experience Fearvana seems too daunting, take other risks. Walk up to a woman, or man, in a bar. Talk to a stranger. Go to a new country. Do anything that scares you. Most importantly, let go of any judgment you have about your fears. Releasing fear's second dart will help you find Fearvana.

One of my clients felt immense fear before traveling by himself to Iceland. He thought of himself as weak for feeling scared of what many people might consider a normal vacation. He believed most people in his shoes would not be afraid. Of course, he could not have known what most others would feel, but we all do that kind of mind reading from time to time.

I worked with him to let go of that second dart and accept his fear as a part of his humanity. His fear of traveling alone to Iceland was no different than my fear of solo mountaineering in Bolivia. To our brains, the experience was the same. It took an equal amount of courage for each of us to push through fear, regardless of what others might think of our pursuits. The only difference between us was I had worked my way up to the point where traveling alone no longer scared me, but solo-mountaineering did.

To expand your comfort zone, Jack Canfield suggests scaling down the perceived risk. "Self-confidence is the result of having successfully survived a risk," he states. Start as small as you need to, but just keep working your way up the ladder of fear. With each new risk you take, you grow in confidence, even if you stick to the same kind of risk.

It doesn't matter how many times we have taken an action, as long as it is still scary and presents some sort of risk, that action can make us stronger. Professional athletes are great examples of this. In an interview with Tiki Barber, former running back of the New York Giants, I asked him if he felt fear before a game. He looked at me as if it was a stupid question. With football players chasing him to pummel him into the ground, he said, "Of course I am afraid."

Before a playoff game, Chris Bosh from the Miami Heat once said, "It's always good to have a healthy dose of fear going into a situation like this. That is a big part of success; you're going to have to play with that fear. It makes you think quicker; it makes you do a little more. It's going to motivate us to play a lot better."

Despite performing at hundreds of concerts, The Boss himself, Bruce Springsteen, openly admitted he stayed near a toilet before a concert because he sometimes vomited. He said, "That's not me being nervous. That's just me getting excited." Excitement and fear are products of the same neurological conditions. The only difference is that we have different beliefs and attitudes about them. The Boss deemed his nausea to be excitement, so he channeled it into mastery. Others might see the nausea as a sign of weakness and ultimately choose not to face the fear head on.

As a child, I used to be scared of everything, so I fled when confronting the terrifying prospect of riding a Ferris wheel. I should have told myself, "This fear is a good thing. This is my mind and body helping me perform better." I just didn't know how to embrace fear back then.

You might notice everyone in the examples above voluntarily placed themselves in a situation warranting fear. They pursued a clearly defined, consciously generated goal. Successful people do not achieve success by accident; they choose to take the risks that lead to their desired results.

It is possible to experience moments when fear, excitement, and risk can present themselves without us consciously choosing them, but Fearvana is not about experiencing fear just to feel afraid. Fearvana is about ascribing purpose or intention to fear. Fearvana is about engaging *your* fears for *your* growth. Fearvana is seeking your own worthy struggle to savor the bliss and success within it and on the other side of it. That means taking 100 percent responsibility for your life and taking your destiny into your own hands.

Action

1. Taking Action toward Fear

This is where the magic happens. In my interview with cave diver Jill Heinerth, she told me, "If you don't chase fear, you spend the rest of your life running from it." Or, as I put it, if you don't seek a worthy struggle, struggle will find you.

Successful people in any walk of life face fear head on and take action in the face of it. Base jumper Jeb Corliss was once asked if he felt fear before leaping off a cliff. He responded, "It's terrifying. What I do is horrifying. As scared as you are standing on the edge of a building and stepping off is exactly how scared I am. I have learned how to control, manipulate, and use that fear to push forward and push through and make my dreams come true." Corliss became a base-jumping icon because he moved into fear, not away from it: "Anytime anything scared me or I felt that sensation of fear, I was drawn to it."

The principles of neuroplasticity apply to the fear response as well. Every time Corliss trained in fear, he taught his brain to embrace fear in the future. Remember, neurons that fire together, wire together. So the more you practice pushing into your fears to experience Fearvana, the better you too will get at handling them the next time they show up.

2. The Disappearance of Self-Consciousness

Your human brain sets you up for Fearvana, but once you reach it, a major part of your brain sits on the sidelines. As Dr. Ilan Goldberg from the Weizmann Institute of Science says, "The term 'losing yourself' receives here a neuronal correlate."

By silencing the chaos of our mind, Fearvana experiences dedicate our cognitive energy to the task in front of us, allowing us to immerse ourselves in the action. "When the climb gets technical, and especially when there is a level of danger, you become utterly present," says climber Margo Talbot. "There's no stress, sometimes even no fear. You literally become simple consciousness."

During this state, only certain parts of the human brain shut off. Other parts are needed to maintain the desired amount of self required to have the focus, presence, and vigilance to move through the action. "Parts of the prefrontal cortex (the human brain) are temporarily deactivating," says neuroscientist and ultra-runner Arne Dietrich. "It's an efficiency exchange. We're trading energy usually used for higher cognitive functions for heightened attention and awareness."

A base jumper needs to be intensely aware of how his body is moving through the air. A one-inch shift in the tilt of the body could result in death. A climber requires full presence to move up the face of the rock without falling. So it is not self or consciousness that is lost, but rather, as Csikszentmihalyi says, there is "only a loss of consciousness of the self."

When consciousness of the self disappears through the deactivation of the human brain, it shuts off our inner critic, freeing us from the burden of processing every moment of our lives. We no longer have to think about how we look, what someone else thinks of us, whether what we said was right or wrong. We stop thinking about the problems of the past and the anxieties of an unknown future. It is a deeply spiritual moment to become what we truly are—human *beings*!

"Suddenly, everything was flowing," Reverend Neil Elliot said about his experience snowboarding. "I was both in and out of time, there and not there. It was just pure being." Through the act of being fully present within an individual moment in time, everything becomes new and exciting. In those brief moments, the feeling of being alive becomes not just blissful but eternal. For if there is no time, how can there be an end to it?

Eckhart Tolle aptly summarizes this quest for timelessness in his bestselling book *The Power of Now*: "The reason why some people love to engage in dangerous activities such as mountain climbing is that it forces them into the now—that intensely alive state that is free of time, free of problems, free of thinking, free of the burden of the personality."

An important point to note here is that this disappearance of self-consciousness comes and goes to a varying degree in Fearvana, depending on the chosen activity. For more cognitively challenging Fearvana experiences, like writing a book, consciousness might be an active part of the process, especially if the book involves a lot of research. There were times when I found myself completely immersed in writing, but often I had to reactivate consciousness to think through the material.

The same thing occurred for me on that mountain in Bolivia when consciousness returned to command my attention once again. This doesn't necessarily mean self-doubt and fear always accompanied the reactivation of my human brain. More often than not, I would go back and forth seamlessly between consciousness and the disappearance of it. This dance between pure, immersive engagement and mindful, deliberate exertion is a spiritual experience that can only be truly understood by leaping into Fearvana. Ultimately, when both consciousness and the loss of it come together, it gives rise to oneness.

3. A Sense of Oneness

Fearvana destroys the natural state of chaos in the brain. When fear shows up, we need all our neuronal resources working together as one to manage the risk that produced the fear. "The brain is always trying to strike a balance between working on different problems at the same time. But there's a mechanism, which under extreme situations allows the brain to take all that processing power and direct it to one thing," says professor of neuroscience Dr. David Eagleman. "The amygdala kicks in and says, okay everybody, you're working for me right now. You have to put yourself into extreme situations to get your brain to do that. When that happens, you unmask its tremendous resources." Fear is the gateway to our greatness.

"It was like I reached a place where clarity and intuition and effort and focus all came together to bring me to a higher level of consciousness," said whitewater kayaker Sam Drevo. "A level where I was no longer me; I was part of the river. It was an amazing experience." When the fear melts away and we flow into Fearvana, all separation disappears. We become connected to everything— to the environment, to the people with us on the ride, and to ourselves.

Once again, we step outside of science and delve into the spiritual. To anyone who has ever experienced it, the feeling of connection and oneness is as objectively real as anything else science can prove. "Again and again waves crashed across the deck until water, air, and iron became one," said Peter Matthiessen of being on a navy ship during a brutal storm in the Pacific. "Overwhelmed, exhausted, I lost my sense of self. The heartbeat I heard was the heart of the world. I breathed with the mighty risings and declines of earth." This sense of unity shows up as a survival mechanism that gives us the strength to endure extreme conditions.

While soloing up that mountain, there was no room for me think about bills or my job. I was forced to become a part of my environment, or I would have paid the ultimate price. A side effect of forming that allegiance with the mountain was that I returned from it with a greater sense of acceptance and humility. With conditions at their worst, I would never stand a chance on that mountain. I am nothing compared to the power of the natural world. How could I then presume to be better than my environment? Or anything and anyone for that matter? I became a part of it all.

Humility is a direct product of oneness. This subsequently creates the space for us to remove our second darts. How can we beat ourselves up when all our weaknesses and strengths are one? I know I am not the fastest or strongest runner, but during my cross-country runs, my weaknesses did not matter. They became a part of who I am. I stopped comparing myself to better runners and reveled in my own progress through each and every painful step.

Through Fearvana, we become one with our own humanity. Self-acceptance and self-confidence are some of the many gifts of Fearvana.

4. A Greater Sense of Control over Your World

As you read this book, you are comfortable in the knowledge that your home will not collapse, a zombie will not break down your door, nor will a meteor fall from the sky. You have some sense of control over your world. The problem in this "normal" world is that control is not exercised or explicitly practiced. However, in a Fearvana-producing environment, control must be practiced to face the dangers and manage the risk.

In 2012, along with five other people, I spent one month dragging a one hundred and ninety-pound sled for three hundred and fifty miles across the second largest ice cap in the world. The environment in Greenland was unforgiving and unpredictable. In fact, just one year after we made the crossing, a British explorer lost his life in a storm just like the ones we experienced.

To survive in such an inhospitable world, we had to take command over it. When brutal storms battered us, we regained control by setting up our tents in a manner that shielded us from possible death. But the reality is we were never in control; Mother Nature was. Although we did not objectively have it, we felt a greater sense of control because we were forced to exercise its muscle every single day.

By silencing the chaos of consciousness, immersing ourselves in the now, and focusing only on our immediate goal, life became incredibly simple. The complexities of excessive choice and the distractions of daily life withered away, giving us complete command over the the world in front of us. "It is not possible to experience a feeling of control unless one is willing to give up the safety of protective routines," says Csikszentmihalyi. "Only when a doubtful outcome is at stake, and one is able to influence that outcome, can a person really know whether she is in control."

I was once almost killed by a falling boulder while ice caving in the Himalayas. I had no way of knowing that boulder would fall, but I prepared for it by learning the environment. I knew the area of greatest danger and made sure to move through it as quickly as possible. A minute later and you would not be reading this book right now.

In nature's unforgiving playground, we learn to mitigate the risks through training. Fear propels us to prepare. As climber Al Alvarez said, "The pleasure of risk is in the control needed to ride it with assurance so that what appears dangerous to the outsider is, to the participant, simply a matter of intelligence, skill, intuition, coordination—in a word, experience."

Elite athletes often respond to the unpredictable events in their environments before they are even aware of their response. A skier in the backcountry turns before consciously recognizing the rock that suddenly appears in front of him. Neuroscience has proven we do this even in our everyday actions, but during intense experiences of Fearvana, acting without awareness becomes more pronounced (and necessary) because of the high level of risk in the environment. That instinct is enhanced through extensive training.

For peak performing athletes, the ability to react to the environment is deeply rooted within their implicit memory. Creativity, resourcefulness, quick thinking, and adaptability have become a part of who they are. They don't know what every situation will look like beforehand, but by training their bodies and their minds in the zone of Fearvana, they have learned how to rapidly react to the unknown.

5. Balancing the Action within the Zone of Fearvana

The zone of Fearvana is the balancing point between the level of fear and the level of training the action demands. The zone of Fearvana exists at just the right intersection between the two. This is what unites flow and deliberate practice within Fearvana and allows the experience of both in one action, as I experienced on the mountain in Bolivia.

"The right dose of fear, of tension and uncertainty can motivate, and allow us to push through expectations. Too much fear is limiting instead of liberating," says Twight. Too much fear and not enough training will flood your brain with excessive adrenaline, often leading to panic.

"I'm the farthest thing from an adrenaline junky," says whitewater kayaker Tao Berman. "I can't stand that feeling. If I'm feeling adrenaline, it means I'm feeling too much fear. It means I haven't done my homework. It means it's time to get out of my boat to reassess." When Berman finds himself outside the zone of Fearvana, he steps out of the situation, goes to his human brain to reassess the situation (he returns to the pre-action phase), and sets up the action again in a manner that places him within his zone. He must either advance his level of training or paddle a safer stretch of water to lower the level of fear.

However, the amount of risk presented by the river doesn't really matter. In any situation, the level of objective risk and the intensity of the fear response do not go hand in hand. Remember, what we think we are afraid of is not always the same as what we feel we are afraid of. Fear from talking to a stranger, jumping out of a plane, running into a burning Humvee, going into a job interview, speaking in public, sitting in front of a computer, or traveling to a new country are all the same to your brain, regardless of the level of risk you might consciously attach to any of those activities.

So the next time you hear someone say things like, "There's nothing to be afraid of," or "Don't be scared," ignore them. That advice is what leads to the downward spiral of second dart syndrome. The client I told you about earlier didn't consciously believe traveling to Iceland presented a real risk, yet his brain produced a strong fear response, for which he then proceeded to beat himself up.

The level of inherent risk to any activity is irrelevant. Some people might need to risk their lives to come alive while others might simply need to risk looking like a fool in front of a few people. All that matters is the activity is one you have consciously chosen in service of your growth and it pushes you beyond your ability to scare you just the right amount.

As neuroscientist James Olds says, "You don't need a giant wave or a big mountain to trigger these responses. The brain's reaction isn't dependent on real, external information. It's reacting to a constellation of inputs from the sensory systems. If you can light up that same constellation—say, replace the novelty found in the natural environment with new routines in daily life—you'll get the dopamine and norepinephrine (adrenaline)." As you keep extending your zone outward, you might find the level of objective risk in the activities you choose increases, as it has for me. Nonetheless, to continue your growth and happiness, focus on maintaining the right balance between your level of fear and your level of training in the activities you pursue.

According to Twight, growth happens when "we pass a threshold of approximately 75 percent of our ability. The closer we get to our potential (in any situation) the more fearful we become. We must operate at this intensity (75 percent) and higher in order to improve. The stress, cognitive, and physical demands below this level aren't great enough to cause change." Too little fear with too much training will lead to an experience without the possibility of true growth mentally, physically, emotionally, or spiritually.

This is not to say such experiences can't be fun. I enjoy going for relaxing hikes with my family, even though they aren't physically or mentally challenging. Of course, there is nothing wrong with experiences that do not push us far enough to reach Fearvana. Fun and pleasure are important and necessary, but a life without fear does not lead to long-term existential enjoyment or growth of the self.

We need the struggle, frustration, and challenge of deliberate practice. Progress does not take place when we live our lives at our limits; it occurs when we play in the space beyond them. "Pushing your own limits is the only way to know yourself and to grow as a human being," states explorer Mike Horn in his book *Conquering the Impossible,* the story of his twelve thousand-mile journey around the Arctic Circle.

So how do you find the ideal fear-to-training ratio that leads to your zone of Fearvana? You experiment! There is no right ratio for any one person at any point in time. It constantly changes. The awareness you gain in the pre-action phase of Fearvana will help you create the right conditions to take action within your zone. "You have to train your body to prepare for the state, you have to train your mind to prepare for the state. You have to know yourself, and your limits, know exactly what you're afraid of and exactly how hard to push past it. That's serious work," says Mike Horn.

Growth is a constant cycle of conscious awareness with automated action. If you don't know yourself well yet, keep practicing some of the self-awareness exercises I have given you. As you keep striking the right balance between awareness and action, the real you will begin to reveal itself. At that point, it becomes much easier to immerse yourself in the fear that leads to Fearvana.

If you are an avid practitioner of Fearvana like Horn, you probably know what your zone is and have no trouble finding it over and over again. For those of you untrained in Fearvana, it will be harder for you to get there initially, and it will mean a great deal of discomfort. Sorry, but there is no way to strengthen your ability to push through fear without feeling it.

The good news is that every time you move into the unknown, your brain releases dopamine in response to the uncertainty. Dopamine makes the experience extremely pleasurable. As Stanford neurologist Robert Sapolsky says, "Maybe it is addictive like nothing else out there." Taking the first step makes it all the more exciting to take the next one and keep expanding your zone.

Once ability reaches a point where fear no longer occurs, as it did for me on that climb with my wife, we must search for the next point of growth. Mountaineers climb tougher mountains. Surfers surf bigger waves. Runners run longer distances. Base jumpers no longer just leap off cliffs; they now fly as close to them as they can. Each of these athletes plans these endeavors with precision. They consciously choose a goal that keeps them within their zone to ensure continued growth.

Successful entrepreneurs do the same. Among his many other endeavors, Sir Richard Branson went from owning a record store to an airline to creating civilian flights to space travel. To reach greater levels of success and performance, it is necessary to keep pedaling on the cycle of awareness combined with action.

Now that we have taken the action, let us return back to awareness in the final phase of Fearvana.

Post-Action

1. Renewal of Self-Consciousness

The post-action phase of Fearvana cements all the lessons we learned within it. It's like hitting the gym. Your muscles don't get stronger when you work out; in training you are tearing them down. Your muscles grow when you are resting and recovering after the workout. Similarly, at the completion of our chosen Fearvana activity, the renewal of self-consciousness fuels our growth, mastery, and progress. "It almost seems that occasionally giving up self-consciousness is necessary for building a strong self-concept . . . Following a flow experience, the organization of the self is more complex than it had been before. It is by becoming increasingly complex that the self might be said to grow," writes Csikszentmihaly. He defines complexity as the "result of two broad psychological processes: differentiation and integration."

What that means is separation and, paradoxically, union. The Fearvana experience separates us from the average individual in that it leads to increased skill, growth of the self, and confidence. At the same time, it leads to the integration of all parts of our self, our environment, and the people around us. When we return from that state of pure being, we attain a new level of consciousness that has been enriched by our accomplishments within Fearvana. We connect with our unique greatness, or as bestselling author Paulo Coelho calls it, our personal legend.

Upon the renewal of my self-consciousness after battling severe heat cramps toward the end of my twenty-eight-mile run across Barbados, I could not have been prouder of my accomplishment. I was alone, so it was not pride in a sense of boasting to others about what I had done. As I placed my head over the toilet seat, ready to throw up from the heat exhaustion, I felt a deep sense of internal pride at having pushed myself beyond my limits to accomplish something I deemed worthwhile. The memory of limping along for those last few miles is a blur, but the reward of battling through that intense suffering will last me a lifetime.

2. Consolidation and Reflection

There are two stages to this final factor. While we reflect on the Fearvana experience consciously, our animal brain works on consolidating the memory of the experience. The dopamine release during Fearvana facilitates the process of learning by increasing our brain's ability to recognize patterns so we can better navigate our way into future unchartered and unpredictable waters.

Exercising our courage in one environment trains our body, mind, and spirit to take control of the only two things we can change in any environment: our attitude or our behavior. Think of every foray into Fearvana as building the muscle of taking 100 percent responsibility for all areas of your life.

Dopamine has also been shown to improve long-term memory and consolidation of that memory. "Dopamine can rescue memories that may have faded otherwise," says neuroscientist Dr. Rumana Chowdhury. Entering your zone of Fearvana trains your brain to face fear and makes it easier to pull that experience from your implicit memory into your explicit one. This not only makes you better at your chosen Fearvana activity, it also adds to the set of references you can use, giving your brain more ammunition to handle everything life throws at you. "Athletes in flow likely gather more relevant data and code it more efficiently," says sports psychologist Michael Gervais. "Having these experiences frequently could significantly shorten the learning curve toward expertise."

So far, all that I have described in the post-action step occurs without your awareness. As your brain goes through this procedure and rewires itself, it can feel like an emotional roller coaster. Once your body and mind settle after an intense experience, it is natural to go through a series of highs or lows.

In the wake of pulling Greg out of a burning Humvee, Dale felt "lost in shock." He began praying to find solace from the horrors of the experience. Someone else might have felt the need to talk to someone about it. There is no right or wrong way to process an experience that pushes you beyond your limits. And, of course, it doesn't have to be something as intense as war either. Any high-stress, fear-inducing event can trigger an emotional reaction upon its completion.

A job interview, exam, business transaction, or athletic competition is often followed by a variety of emotions. You might do something, like look back at the event trying to figure out what went well and what went badly. Or you might feel happy, sad, scared, relieved, grateful, frustrated, angry—you name it. The assortment of emotionally charged responses available to you help you find the value and meaning of your experience, which flows directly toward the conscious part of this final factor: reflection.

The conscious process of reflection can involve sitting down with a pen and paper to document the action or simply thinking about how it aligned with your purpose in life. Either way, it requires focus and effort to analyze the experience and take from it whatever will serve your growth. For example, the factors for Fearvana listed in the action phase could apply to something like gambling. However, upon reflection, you might deem such an activity unworthy of your growth and not worth replicating in the future.

Reflection helps us consciously implant the value of any Fearvana experience. It also helps us become aware of our zone of Fearvana, giving us direction for how to replicate that experience and enhance it again in the future. "Activity and reflection should ideally complement and support each other. Action by itself is blind, reflection impotent," writes Csikszentmihalyi. The three phases of this fear to Fearvana system represent a process I call the action/awareness cycle. Using our awareness to take action and then reflect on the action is how we walk our path of Fearvana.

Chapter 9

THE PATH OF FEARVANA

At the age of sixteen, I found myself lost in a dark, dangerous world. For over a year and a half, I squandered away my life with drugs. My friends and I were so heavily immersed in a lifestyle of self-destruction that two of them died as a result of it. I too was headed down that path.

Drugs were my chosen method to test my limits. From snorting lines of cocaine to taking eighteen hits of acid, I was destined for a short, meaningless life. Until one day when I saw the movie *Black Hawk Down*. Witnessing the courage of men sacrificing their lives for their fellow human beings made me wonder if I would be able to do the same thing. I didn't have an answer, so I decided to find out.

After watching the movie, I read the book about the true story it was based on. I then read as many books on war and the military as I could find. That movie was a spark that set in motion a series of events which led to me quitting drugs almost overnight and enlisting in the United States Marine Corps. I did so despite two doctors telling me boot camp would kill me because of my blood disorder, thalassemia.

With a firm conviction about the path I had chosen and an unwavering belief in my ability to walk it, on July 12, 2004, I set foot on the infamous yellow footprints at the Marine Corps Recruit Depot in San Diego, California. I not only survived boot camp but graduated infantry school as the honor graduate in my platoon. Three and a half years later, halfway through a deployment in Iraq, I had no idea what I would do with the rest of my life, if and when I made it home. One day, my friend lent me his copy of *Ultramarathon Man* by Dean Karnazes. Once again, a spark lit.

Inspired by the book, I used whatever time and space I had to build myself up as a runner. Within the protection of "the wire," which consisted of large blocks filled with sand, we could move freely without wearing flak jackets or Kevlar helmets. I made sure to take advantage of that luxury. Sometimes, I woke up early to run before our mission. Other days, I ran in the afternoon between missions. On a rare few occasions, I managed to get in up to three hours of running, which meant about one hundred and fifty laps around our little home in Iraq.

That spark has now become a raging fire. I am now on a seemingly impossible journey to run from border to border across every country in the world.

How Does Success Start?

In November 2005, *Forbes* labeled Roger Bannister's feat of breaking the four-minute mile the "greatest athletic achievement of the 20th century." By proving it could be done, Bannister implanted a new belief in the minds of runners everywhere. "There was a mystique, a belief that it couldn't be done, but I think it was more of a psychological barrier than a physical barrier," Bannister said. His accomplishment triggered seventeen other runners to replicate his achievement within three years, the first of which occurred within forty-six days. Greatness in every endeavor seems to follow a similar pattern.

In 1998, Se Ri Pak became the first South Korean woman to win the LPGA tour. Her victory became a spark that led to forty-five South Korean women collectively winning almost one-third of the LPGA events within ten years. When Anna Kournikova reached the Wimbledon semifinals in 1998, within less than ten years, five Russian women were in the top ten spots of the Women's

Tennis Association (WTA) rankings. One spark has the power to trigger a surge of subsequent success.

These spark moments are not just the result of something as grand as someone winning a major tournament. After struggling at basketball camp, Shaquille O'Neal came home and told his mom he wasn't sure if he wanted to play basketball anymore. When his mom pushed him to try harder, he said, "I can't do that right now; maybe later." His mother told him, "Later doesn't always come to everybody." Such seemingly trivial, everyday moments have the power to transform an ordinary person into a legend. "That got to me. Those words snapped me into reality and gave me a plan," O'Neal said. "You work hard now. You don't wait. If you're lazy or you sit back and you don't want to excel, you'll get nothing. If you work hard enough, you'll be given what you deserve."

Despite his fear of water, Michael Phelps began swimming because of his two older sisters. They were the spark that first got him in the water. Then in 1996, Phelps witnessed Tom Malchow and Tom Dolan at the Atlanta Summer Games. Watching them compete ignited a fire that turned him into the most decorated Olympian in history. When Michael Jordan was rejected from his high school varsity team in favor of his best friend Leroy Smith, he promised himself something like that would never happen again. The disappointment and shock of that moment became his spark. No matter where you find it, success follows a spark event.

During his research, *New York Times* bestselling author Daniel Coyle visited numerous "talent hotbeds" around the world to find a common thread on what it takes to achieve greatness. These were places that seemed to consistently produce peak performers in a variety of endeavors. He found externally generated spark moments, what he refers to as "ignition," are the triggers that lead to long-term success. "We usually think of passion as an inner quality. But the more I visited hotbeds, the more I saw it as something that came first from the outside world," he writes in his book *The Talent Code*.

It would make sense that inspiration first comes from the outside world. No one is born with an innate desire to become an Olympic athlete, a Hollywood actor, a successful entrepreneur, or anything at all. We learn to become these things from the environment around us. "Our environment is the most potent triggering mechanism in our lives," writes Marshall Goldsmith in his bestselling book *Triggers.* Our referential brain constantly searches for references to help us choose the right path from the moment we set foot on this earth.

Even those believed to be "child prodigies" were not born with success in a particular pursuit encoded into their DNA. Tiger Woods was raised by a father who had a single-figure handicap in golf, which is no easy feat to achieve. When Woods was nine months old, his father cut the top of an old golf club and gave it to him to start practicing with. By the time he was eighteen months, his father was taking him to the driving range. Regardless of his genetic conditioning, Woods was not born a golf champion; he was made into one.

Even one of the most famous child prodigies in human history, Wolfgang Amadeus Mozart, wasn't born a genius. Like Woods, it was his father, Leopold Mozart, who helped him become one. It has been estimated that by his sixth birthday, Mozart had already spent thirty-five hundred hours studying and practicing music. A spark followed by painstaking practice produced results, not some genetically ingrained sense of purpose.

The inherent problem this presents is that a spark is generated externally. This resigns control of our destiny to a world we have no control over. Both the spark moments in my life, *Black Hawk Down* and *Ultramarathon Man*, took place by accident. I did not seek them out; they just came my way. I simply took hold of the hook and worked hard as I went along for the ride, as did Forrest when he overcame his brain injury. He happened to come across a TV show discussing Jack Canfield's book *The Success Principles.* That became his spark, which then triggered countless hours of effort to rebuild his brain.

So if you don't have parents who pushed you into golf, if you didn't happen to read a book or watch a movie that transformed your life, if you haven't found your spark yet, what can you do? You can consciously generate your spark.

How to Create and Ignite Your Spark

"Work gives you meaning and purpose and life is empty without it."
—**Stephen Hawking**

The first step in generating a spark is to become more conscious about the environmental cues that shape your decisions, your perspective on life, and your own abilities. You do have the power to choose where those cues come from and what you want to abide by. If you don't take control of your environment, it will control you.

If you want to become an athlete, surround yourself with stories and studies on athletic performance. If you want to be a writer, learn more about how writers succeed. If you want to be an entrepreneur, study the art of entrepreneurship. We always learn from others, so if you don't know how to do something, find someone who has done it and copy them. Borrow the experience of others to fuel your own growth. That's exactly what I have done throughout this book.

However, when modeling mastery, don't just focus on what people did or how they did it. If you really want to learn from someone, dig deep to understand the belief systems that drive their behavior. Ask them questions like, "What are your beliefs about the world? How do you see yourself? What do you believe is true about this thing I want to learn from you?"

When the president of Advanced Behavioral Modeling Inc., Wyatt Woodsmall, trained a group of army recruits how to shoot, he asked a variety of shooters, "What do you believe about the gun?" Novices believed things like guns are dangerous; it might hurt me; or it will backfire, whereas experts believed things like the gun is safe; the gun is my friend; and it will protect me. Woodsmall would then have the new recruits close their eyes, get into relaxed state, and repeat affirmations of these new beliefs to start incorporating them into their subconscious.

Human beings do not do things that violate their model of the world and their self-identity. The animal brain is in charge, so we want to know the subconscious forces governing the animal brain of the people we seek to emulate. Start with the why; then get into the how. Going deeper into the psyche of peak performers will radically improve your ability to reach their level of mastery or exceed it.

Whatever you want to be, do, or have, find references to validate your ability to turn that dream into a reality. Seek out references that will inspire you and implant a belief in the possibility of the life you want to live. Your referential brain does this subconsciously anyway, just become more proactive about it.

Beliefs set the foundations from which we act, consciously and subconsciously. But, the process for change is not linear. It's like the chicken and the egg; it isn't always clear what comes before the other. Just as you can change your behavior by changing your beliefs, you can also change your beliefs by changing your behavior. That is why this entire system of growth is called the action/awareness *cycle*! Which part of the cycle you choose to enter will often vary, depending on the situation.

In his late twenties, Sir Richard Branson found himself stuck in Puerto Rico on his way to the British Virgin Islands. "They didn't have enough passengers to warrant the flight, so they cancelled the flight," he explained in a documentary about how he started Virgin Airlines. "I had a beautiful lady waiting for me in BVI and I hired a plane and borrowed a blackboard and as a joke I wrote Virgin Airlines on the top of the blackboard, $39 one way to BVI. I went out round all the passengers who had been bumped and I filled up my first plane."

At the time, he didn't have enough money to charter that plane, yet he took the risk and then found a way to make his money back. He took action first and then used the awareness from that experience to start Virgin Airlines. If you constantly get paralyzed by fear or perfectionism, you might need to act without thinking. Telling yourself, in the wise words of Sir Richard Branson, "Screw it; let's do it." In other situations, like quitting a job to start a business, especially when you have a family to provide for, you might need to exercise your awareness first, before taking action.

Either way, though, you will never feel 100 percent ready to set foot into the unknown. How could you? It is the unknown, after all. Whatever you want to do, whether it be running a marathon, building a business, or jumping out of a plane, at some point you will need to take action—even if you don't feel entirely ready for it. Successful people feel the fear, leap into it, and then fight hard against any obstacle that comes their way.

However, you do want to be careful when, where, and why you start something. The more you start things without finishing them, the more you ingrain a belief that you are not a finisher. If you plan a five-mile run and only do four, you will start to believe you are someone who doesn't follow through. That belief will carve its way into your self-identity and show up in all areas of your life. Ignore the often-preached idea to just jump into things because "it can't hurt to try." It most definitely can hurt. You do not want to form a definition of yourself that you are a failure who doesn't finish what he starts.

To avoid jumping into something blindly, be careful of "shiny light syndrome." Today we are bombarded by so much information that once we pick a path, it becomes easy to let it go and jump to another one. We are surrounded by so many possible sparks, it becomes hard to figure out which is the one that will turn into a fire.

Evading this demon called "shiny light syndrome" requires awareness and immense clarity of purpose, like setting the date by when you want to accomplish a specific result. In Sir Richard Branson's case, his purpose was a beautiful woman waiting for him, and he wanted to reach her that same day. For you, it could be losing ten pounds in sixty days. Whatever it is, having clarity of the outcome will help you stay on track and prevent you from jumping into something else.

Another demon to watch out for as you embark upon your path of Fearvana is the "comparison demon." This is an especially challenging obstacle in the social media world. When we see others who are on the same path as us, but much further along, we often beat ourselves up or start to believe we can never do it. When the "comparison demon" shows up, remind yourself you are comparing your present, relatively unaccomplished self to their best self. Everyone starts at point zero; it took work to get to where they are now. If you want to be inspired by someone else, focus on the work they put in, not just the results they achieved.

How you choose to interpret your environmental cues will determine how they affect you. As long as you keep practicing the art of control over your consciousness, you will be able to recognize and leverage the spark moments happening around you, like a charter plane available for hire. The next step in your evolution will then be to get out there and create such moments.

My path in life has changed a great deal in my thirty years. Initially, sparks came to me by accident, but as I learned more about what it takes to realize my legend, I consciously generated them. My goal to run across every country in the world came about as a result of consciously searching for something seemingly impossible to engage in as a lifelong project. I found my spark in an ultra-runner named Pat Farmer. He ran from the North Pole to the South Pole in ten and a half months, averaging about two marathons a day, without a single day off. His incredible feat showed me the word *impossible* meant nothing.

I don't know when I will complete my mission to run across every country, or even if I will, but it doesn't matter. Having that goal serves as a compass to keep me focused and hold chaos at bay. But the real reward comes from each step of the journey, from my training, to the cross country runs, to the people I meet, to the joy of coming back home to my wife and family—all of it.

Clarity on my path of Fearvana has not only given me enjoyment and meaning in the ongoing stream of experience, but it has also liberated me from the environmental cues thrust upon me by society, including the ones where people tell me what I am doing is stupid, impossible, or insane. We combat the fear that is exploited by others when we actively seek it out in service of our growth. With our compass set, we leave no room for others to manipulate our animal brain.

Peak performers in any field stay on their path because they have an internally driven clarity of purpose that gives them the means to separate themselves from their environment. Regardless of the economy, the political climate, or how they are judged by the world around them, they keep moving forward with unwavering tenacity.

Once you create your own path of Fearvana, the one you want to walk for the rest of your life, let it consume you, and you too will find the courage to keep putting one foot in front of the other.

The Myth about Work/Life Balance

"Rowing for me had become a painful, most persistent itch that had gradually taken over my life."

—**Brad Alan Lewis**

In India, many parents put immense pressure on their children to get good grades so they can get into a good college. This occurs at a great cost to the students' well-being. In 2013, 2,471 students committed suicide because of "failure in examination." That's almost seven young people taking their own life every day.

All over the world, people often choose paths designated to them. If you want fear to work for you, you need to know exactly what lies waiting for you on the other side of that fear and why you want to take the trip, separate from what anyone else tells you. The secret to repeatedly engaging in the struggle inherent to the experience of Fearvana is to have something worthy to struggle for, something to focus your consciousness on long-term to silence the inner chaos and give meaning to the inevitable lows and highs of life. As Csikszentmihalyi states, "When a person's psychic energy coalesces into a life theme, consciousness achieves harmony."

Find one purpose for your existence that everything falls into. Your path of Fearvana is the plan for your life that sucks everything into it. I like to visualize a black hole actually sucking in everything, even light, into its all-powerful grasp. From that place, I find it nearly impossible to waste the moments I have left. That image forces me to become consumed by the vision and goals I have for my life.

As Bobby Maximus says, "If you truly want to be great at something, you need to be obsessed with it. You need to want it more than anything else in your life. It needs to be the first thing on your mind when you wake up and the last thing on your mind when you go to sleep at night. You need to dream about it. When you truly want to be great, there is no sacrifice too big to make. You walk, talk, breathe, and live your goal. Let obsession light the way on your journey."

Forget about the often-preached concept of work/life balance. That is not how rower Brad Alan Lewis won the Olympic gold. He gave every bit of himself to his one pursuit and was even willing to die for it. And why not? "When you expend an immeasurable amount of effort over thirteen years, all toward a solitary goal, then a commitment to die trying is a natural evolution," he said. That kind of all-consuming fire is the difference between mastery and mediocrity. Greatness demands obsession, not balance. When you find the fuel that drives you, everything will flow directly into the black hole that is your path of Fearvana: your work, your life, your environment, your relationships, your playtime—all of it.

My work is a part of who I am. I talk to my wife about it all the time. I love what I do, so it is only natural for me to bring my work home with me. Similarly, I bring my home life to work. I share personal examples with my clients all the time, just as I have done with you in this book. Collectively, all parts of my life integrate with each other to form a sum greater than its parts. As Sir Richard Branson said, "I don't think of work as work and play as play. It's all living."

Peak performers in any area integrate all parts of their life onto their path of Fearvana. This doesn't mean you have to be "on" all the time. We need time off to relax; we need to spend time with friends and family; we need time to enjoy our favorite hobbies; those activities are all important. The point is simply to give the entirety of your life a clear and meaningful purpose, so even when you do "turn off," it is a conscious decision in service of the overall advancement of your life.

The growth of your true self transcends the limitations of either your work life or your personal life; it encompasses both of them. With the kind of clarity that pulls all parts of your life into one cohesive whole, there will be no room for the demons to enter. This will make you excited to get out of bed in the morning and feel alive every day as you walk your path.

Of all the strategies I used to beat alcohol, becoming intensely clear about my destiny was perhaps the most effective one. I surrounded myself by reminders of the person I wanted to become. I watched ultrarunning and polar exploration documentaries. I put pictures and quotes up on a vision board. I primed my subconscious with only the messages that advanced me toward my goals. I did everything I could to keep my goals, and only my goals, in my mind. I then planned out my life and tracked every moment of it in preparation for a specific result. I immersed myself in a worthy struggle that took over the void once filled by alcohol.

As you might be able to tell, I prefer to push my limits in various pursuits, rather than devoting the entirety of my being to one. I have consciously chosen to engage multiple fears across multiple terrains and activities. But everything I do falls within the umbrella of my life purpose. My path of Fearvana is to unleash the limitlessness of the human potential in self and others to create enduring peace, one community at a time. With this in mind, everything I do is a step toward or away from that mission.

There are, of course, many others who choose to commit themselves entirely to one pursuit and strive to become the best in their chosen art form. For Reinhold Messner, his path of Fearvana involved mountaineering. For Jeb Corliss, it was base-jumping. For Steve Jobs, as he described in his original mission statement for Apple, it was "to make a contribution to the world by making tools for the mind that advance humankind." For Stephen Curry, it was basketball.

For you, it might be something else. It might be running, painting, singing, acting, skiing, knitting, or any number of activities. It doesn't matter which path you choose; all that matters is that it is *yours* and you are clear on what it looks like. Clarity and a sense of ownership about your path will help you commit your effort to the struggle. It will also help you find the bliss that lies within that struggle and on the other side of it. As President Roosevelt once said, "The best prize that life has to offer is the chance to work hard at work worth doing."

Training Exercise

"Don't follow your passions, follow your effort. It will lead you to your passions and to success, however you define it."

—**Mark Cuban**

There is a lot of information available on how to find your path in life. You will often hear people say things like, follow your passion and the money will come. The debate on whether or not this is good advice rages on all over the Internet. Based on my research on this subject, my personal experience, and the people I have worked with, I have come to the conclusion that follow your passion is not the best advice. This becomes increasingly true the further away you are from being in a stable life position.

Why? For one, we generally have a lot of passions. We can't follow all of them. If and when we try, we end up bouncing around from one to the other, as I myself have done in the past. The paradox of choice also plays a role here. There is an infinite supply of things to do in one lifetime. Of course, we will be confused as to which one of those options is the right one for us.

Secondly, not all of these passions (if any) are going to pay the bills and/or guide us to the lifestyle we want. Sure, there are people making a living in almost any and every endeavor you can think off. But as bestselling author of *So Good They Can't Ignore You,* Dr. Cal Newport, found in his research, few people turn a passion into a career. Instead, most people choose a path, and their passion for it surfaces as they walk it. As Newport advises, "Don't set out to discover passion. Instead, set out to develop it."

Most of us have no idea who we are or what we are passionate about (which is why the foundation for this book is self-awareness). How could we possibly know what makes us come alive without a set of pre-existing references in our brain connecting bliss to that activity? Life experience establishes those references. Inevitably, then, passion is almost always preceded by effort and exploration.

"When I first entered the pool, I was afraid," said Michael Phelps. "The more time I spent in the water, the stronger I became, and my passion for the sport grew." When we get good at something, we then start to enjoy it. I used to *hate* long-distance running. Now I am planning on running across every country in the world. I figured out my passion through a very convoluted path of self-exploration in multiple endeavors.

As *New York Times* bestselling author Seth Godin says, it "isn't about waiting for the right answer, because there is no right answer. There are challenges we can sign up for and emotions we can experience." Anything we do will always be new to us initially, so fear is the precursor to skill, which is the precursor to passion.

The path of Fearvana will help you find the fears that lead to *your* growth and *your* happiness. We all have our own worthy struggle to follow. Here is a training exercise to help you find that worthy struggle, which will lead you to your passion:

Step 1: Write down everything you don't want to be a part of your ideal lifestyle. Remember, we do more to avoid pain than to gain pleasure. Generally speaking, because of negativity bias, we are also more aware about the things that cause us pain than the things that cause us pleasure. So we are going to start with what we don't want.

Step 2: Write down what you do want your ideal lifestyle to look like. If you have gotten this far into the book, you are obviously committed to improving your life, so you should have enough references to give you some idea of what you want this lifestyle to look like. To help you get more clarity, here are some possible questions to ask:

- What do people currently ask you to help with?
- What do you want your tombstone to say about you?
- Who inspires you and why?
- What is it about his or her lifestyle that inspires you?
- What would you like your day to look like?
- How flexible is your daily schedule?
- How much time would you like to spend with family? Friends?
- How much of an impact would you like to make in the lives of others?
- What kind of work would you be doing?
- How important is it for you to be known for your work?
- Is fame and exposure of value to you?
- Where would you be living?
- How intense is the work you are doing?
- How much time would you like to spend working?

Step 3: Seek out references from your own life and others to find a potential "job" that triggers a spark moment within you. If nothing stands out yet, don't worry; that's perfectly normal. In that case, use the references to create a list of three to five possible "jobs" that meet your lifestyle criteria in steps 1 and 2. I put jobs in quotations, because job doesn't have to mean what is traditionally considered a job. Mountain climbing or base jumping could be the "job" you choose.

Step 4: Choose one job from that list you believe would most likely keep you within your zone of Fearvana, at least in the early stages.

Step 5: Keep seeking out references from your own life and others' lives to find out what it takes to become successful at that job. In this step, I also encourage you to reach out to people directly to get their help. Use the guidelines mentioned earlier in this chapter during your outreach.

Step 6: Follow the steps in the fear to Fearvana system outlined in chapter 8 to become a master at your chosen endeavor.

Step 7: Use the mastery you have developed from your own life and others' experience to create the lifestyle you outlined for yourself in steps 1 and 2.

Step 8: Keep expanding your zone of Fearvana to ensure continued growth in whatever form you choose: spiritual, financial, physical, mental, or contributional. The next section will help with this step.

Step 9: Be open and willing to see where life takes you.

At this point, you should have your basic needs met, which in the modern world essentially means you are financially stable. Once you are living the lifestyle you have chosen for yourself, this is the time to potentially explore new passions. Now you get to choose how, when, and where you want to continue experiencing Fearvana and reach your next level of growth.

Follow your passion is not all-around bad advice. With your needs met, you then have the luxury to do what you want, when you want. That is when "follow your passion" becomes a valuable piece of advice.

For example, after becoming a bestselling author, Jack Canfield did an exercise to figure out his passions and the degree to which he was living them. He discovered he wasn't fully living his passion of connecting with people in his industry to support each other in making an impact on the world. That realization inspired him to start the Transformational Leadership Council, a gathering that unites "leaders in the field of personal transformation."

For Canfield, following his passion led to something remarkable. It pushed him into something he had never done before at a time in his life when he had the knowledge, ability, skills, and resources to venture down that particular path. He developed all those things through years of hard work to first establish himself as a leader in the personal development industry. Then he began exploring new ways to live his passions and continue growing.

The exercise Canfield used to become more aware of his passions was the Passion Test by Chris and Janet Atwood. I highly recommend that exercise when you are ready for it. Check out www.fearvana.com/resources to learn how to conduct your own Passion Test.

Unless you are at the point where you feel you have the freedom to explore your passions, I DO NOT recommend doing the test right now. It will distract you from the immense hard work it will take to develop your passions by first figuring out what your path of Fearvana looks like.

If you are one of the extremely rare few who knows exactly what you are passionate about before having walked on that path, good for you. Use everything you have learned and will learn in the next section to live that passion.

Section 3

AWAKENING

"That is the real spiritual awakening, when something emerges from within you that is deeper than who you thought you were. So, the person is still there, but one could almost say that something more powerful shines through the person."

—Eckhart Tolle

We don't climb mountains to reach their summits. We climb to reach the summits within ourselves. An awakening is when we arrive at our next summit.

In section 1 we got to know the tools of the trade to climb the highest mountains, or achieve mastery over ourselves. In section 2, we learned the art of Fearvana, to take action and climb to the summit. Now we are going to figure out what it takes to keep coming back so we never stop attaining our next awakening.

This section is about mastering life in that space between summits and finding the strength to reengage your fears when you become a little too acclimatized to your comfort zone. Whether you are a pro athlete or someone who is struggling to make it to the gym more than once a week, you will find yourself in that space. Often it will be much harder to navigate than the climb itself.

As General Manager of Gym Jones and one of the 100 fittest people in the world as profiled by *Men's Health,* Bobby Maximus says, "The training is the easy part. What happens the other 22 hours of the day is where the battle will be won or lost. Remember that when you leave the gym the real work begins." Let's start the real work.

Chapter 10

THE GIFT OF SUFFERING

"If your bliss is just fun and excitement, then you are on the wrong path. Sometimes pain is bliss."

—Joseph Campbell

As I changed the battery of my torch, my stiff and tired hands refused to work properly. I dropped the battery condemning us to total darkness. We dug out a shallow cave and hid ourselves in it, shivering from cold and tiredness, cowering close to each other to await dawn. I could feel and see that we were at our physical limits," wrote Jerzy Kukuczka on the night after reaching the summit of K2.

Mountaineers call K2 the savage mountain. It is the second highest mountain in the world and often considered the most dangerous. Its treacherous faces take the lives of one in every four climbers audacious enough to set foot on its summit. Mount Everest, on the other hand, takes the life of one out of every ten climbers who stand on the roof of the world. Even the "easiest" route up K2 is more dangerous than the one on Everest. Yet, for some bold explorers, the easiest route is not enough.

"I have something inside me that makes me have no interest in playing for low stakes. For me it is the high bid or nothing. That's what fires me," said Kukuczka. In 1986, Jerzy Kukuczka and Tadeusz Piotrowski reached the summit of K2 via the South Face, or what is known today as the Polish line. The route is so avalanche prone and requires such intense technical climbing at high altitudes that, to this day, no one has repeated the accomplishment. Mountaineers call it "the hallmark of suicidal excellence."

Both Kukuczka and Piotrowski continue to be immortalized by the mountaineering community for their feats of daring. Piotrowski is known by climbers as "the finest winter mountaineer of his day." After Messner, Kukuczka was the second man to climb all fourteen of the eight thousand peaks. His feat was outstanding in that he did it in eight years, created ten new routes, and summited four of these peaks in winter. The audacity of entering the Himalayas, let alone climbing an 8,000-meter peak in winter cannot be understated. Imagine hurricane-force winds, avalanches, the constant risk of frostbite, and a horrific unearthly cold that *National Geographic* called "the angel of death." It is by far one of the most hostile environments on the face of this planet. Yet, Kukuczka and Piotrowski dared to venture into it over and over again.

To make things worse, Kukuczka's body was not suited for high-altitude mountaineering. In Denali, he was brought to his knees by altitude sickness at just 14,700 feet. Yet he did not give up. He fought his way up the remaining 5,610 feet to the summit through sheer willpower. Having personally experienced the headaches brought on by altitude sickness, I can't even fathom what it would take to keep climbing through such debilitating pain for over 5,000 feet.

It wasn't just Denali that Kukuczka struggled with either. On almost every climb, he found it difficult to acclimatize to the low oxygen pressure at high altitudes and was often much slower to do so than his climbing partners. He never let that come in his way. He compensated for his lack of physical prowess with an iron will and superhuman ability to withstand suffering. His climbing partner called him "a psychological rhinoceros" who was "unequaled in his ability to suffer."

Suffer Now, Savor the Sweet Taste of Glory Later

"A gem cannot be polished without friction, nor a man perfected without trials."

—**Seneca**

Life in the Himalayas is what psychologist Mark Leary calls an immediate-return environment. When Kukuczka and Piotrowski faced a storm, they felt despair. When the storm passed, the despair was alleviated. That is the same environment our ancient ancestors lived in. If they saw a predator, they felt fear, ran away, and the fear was gone. In such environments, stressors are short term because problems are as well. Stress is immediately relieved when the obstacle is overcome.

Today we live in a delayed-return environment. Stressors are no longer resolved nor rewards received from immediately overcoming an obstacle. The choice to study hard and get an A on an exam does not lead to the reward of getting into a good college until years later, if at all. From early on, we learn to live entirely in the future, for that is where we believe our rewards lie. We worry about getting good grades, so we can get into a good college, so we can get a good job, so we can save for retirement.

Yet, we don't know if any of these achievements will actually lead to the next desired result. I spent years and incredible amounts of painstaking effort researching and writing this book, but I have no idea if this work will pay off. In a delayed-return environment, the challenges we face cannot be overcome in one moment of spirited action. I don't mean to make it sound all bad. In a delayed-return environment, we get to explore and solve more long-term, complex challenges than we could if survival was our only concern.

The paradox of this current reality, however, is that we also live in a culture of instant gratification, which has taught us getting what we want is easy. We can order anything we want with one click. We can watch movies instantly on our phones. We can find the answer to any question within seconds. This makes living in a delayed-return environment all the more difficult. We want results instantly, and we get those results when it comes to the little things, but that's not how results are attained for anything worthwhile in life. Earning millions of dollars takes extraordinary effort. Creating a successful relationship is hard. Losing weight to get those six-pack abs takes long hours of suffering in the gym. No great success comes without the beauty of the suffering required to attain it. "A life without anxiety, frustration, competition, and challenge is not the good life," states Dr. Martin Seligman, a leading researcher in the positive psychology movement.

So, how can our brain get the immediate return it desires while committing to the struggle involved in earning the reward from the delayed return? To answer that question, let's look at colonoscopies.

What a Colonoscopy Teaches Us about Happiness

Thankfully, a colonoscopy is no longer a painful experience, but in the 1990s, that wasn't the case. Back then, a study was conducted where patients were asked to rate their pain on a scale of one to ten every sixty seconds while undergoing the colonoscopy. The graph below shows the level of pain two different patients experienced.

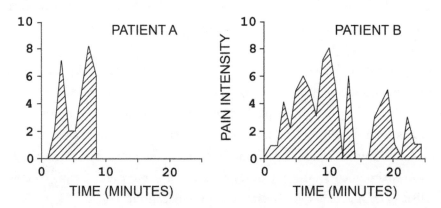

Which patient do you think suffered more? Clearly patient B.

After the procedure, each patient was then asked how much he thought he suffered and "how bad was the whole thing in total." Patient B's colonoscopy was about fifteen minutes longer, and he experienced more pain than patient A. Yet, he had a much better memory of the event because his colonoscopy ended with a low pain rating.

In a fascinating addition to the experiment, researchers extended the colonoscopy of patient A, but made sure not to move the tube around too much. This extended the amount of pain he suffered but reduced the level of pain during that extra time. Patient A then had a worse experience, as he suffered more and for longer, but his memory of it was far better because he now walked away from the procedure with a better ending to the story.

So what does this mean? Within each of us lie two selves: the experiencing self and the remembering self. The experiencing self is the one that lives its life from moment to moment. It is the one reading this line. According to Dr. Kahneman, that moment, what he calls the psychological present, lasts three seconds. The experiencing self lives its entire life within these three seconds. The remembering self is the one that looks back on our lives to find the meaning in it. It is the one who will tell the story of what reading this book was like.

In the case of the two patients undergoing the colonoscopy, there was a clear conflict between these two selves. The experiencing self was better off for patient A, but the remembering self was worse. For patient B, it was just the opposite.

Hypothetically, lets imagine both patients went to another doctor, and the procedure was reversed so that patient A's experiencing self suffered more, but the remembering self suffered less. Which doctor do you think each patient would have chosen for their next colonoscopy? Patient A would choose this hypothetical doctor while patient B would have chosen the real one.

Why? Because "we actually don't choose between experiences, we choose between memories of experiences. When we think about the future, we don't think of our future normally as experiences. We think of our future as anticipated memories," states Kahneman. "You can look at this as a tyranny of the remembering self, and you can think of the remembering self sort of dragging the experiencing self through experiences that the experiencing self doesn't need."

The remembering self is the one that makes decisions for our present self and our future self, based on the stories or references it has from our past. The experiencing self has no say in the matter. The remembering self uses our memories to choose how we live our lives, from the vacations we take, to the jobs we want, to the relationships we seek out.

As we learned from the colonoscopy, story endings have a lot of power. If you have a bad breakup to a relationship, the ending is what will stand out about that entire relationship. It will even shape your decision on the next person you date. The same applies to the last place you worked or anything else for that matter. The remembering self has a lot of power.

On the other hand, the experiencing self disappears after just three seconds. But it doesn't feel right to simply ignore it, does it? Every three second chunk of time must be treasured, for what takes place within those moments is our life.

Why Suffer?

> *"I firmly believe that any man's finest hour, the greatest fulfillment of all that he holds dear, is that moment when he has worked his heart out in a good cause and lies exhausted on the field of battle—victorious."*
>
> **—Vince Lombardi**

There is a slight twist to this story about our two selves. Just as the remembering self drags the experiencing self into new experiences, the experiencing self has the power to alter the stories told by the remembering self.

Remember, our memories lie to us. They tell us a story of the past that changes every time we go back in time. The remembering self is constantly making up "reality" based on what the experiencing self is feeling at the time the remembering self is activated. If the experiencing self is sad, it will cause the remembering self to attach sadness to the memory it brings back into the present. Both selves work hand in hand to shape each other, but clearly they are not always on the same page.

Seeking out a *worthy* struggle is the key to uniting both our selves. It is how we make both of them happy. For the remembering self, that struggle will always be worth it. It is fairly obvious you can't achieve a result unless you work hard to attain it. Success does not happen by accident. Inevitably, then, through Fearvana, the remembering self attains bliss by looking back on a life to tell stories filled with success. For the experiencing self, Fearvana opens us up to the neurological, psychological, and spiritual possibilities described in chapter 8. By gifting us suffering, Fearvana also reveals to us the superhuman courage we all have to overcome that suffering.

"Little by little, I was beginning to get used to the suffering that this remarkable journey was forcing me to endure, to the point that I was starting to enjoy the sensation of overcoming the pain a little more each day," wrote Mike Horn during his 27-month, 12,000-mile journey around the Arctic Circle. When the struggle to success is one we consciously choose, one that is in line with who we want to be, it makes the pain of that struggle beautiful and joyful. Fearvana combines suffering and bliss into a transformative, life-altering moment.

"Contrary to what we usually believe, the best moments in our lives are not the passive, receptive, relaxing times, although such experiences can also be enjoyable if we have worked hard to attain them," concluded Csikszentmihaly, based on his research on happiness. "The best moments usually occur when a person's body or mind is stretched to its limits in a voluntary effort to accomplish something difficult and worthwhile." This is not something that can be fully grasped by reading a book. It is in the experience of seeking out a worthy suffering that we truly find the gift in it. Such moments are not always pleasant in the traditional sense of how we define pleasure, but the joy comes from becoming that pain as it becomes one with us. Unending bliss awaits us within the simplicity of going to war with ourselves.

This doesn't mean taking on suffering for the sake of suffering. Don't take a knife and start cutting yourself, as I often did during my drug phase in high school. There was no virtue in that pain. Only when we take on a worthy struggle with the awareness it is in service of our growth and our happiness does the struggle become pleasant.

As seven-time Mr. Olympia winner Arnold Schwarzenegger said, "Pain makes me grow. Growing is what I want. Therefore, for me pain is pleasure. And so when I am experiencing pain I'm in heaven . . . I like pain for a particular reason. I don't like needles stuck in my arm. But I do like the pain that is necessary to be a champion."

Top athletes don't see a hard session in the gym or the years it takes to achieve mastery as suffering to be avoided. They welcome that pain. All peak performers abide by the growth mindset that enduring such pain demands. They see suffering, whether it comes in the form of practice or stress or failure or fear as access points to their greatness.

Tennis champion Monica Seles said, "I just love to practice and drill." Serena Williams further echoes the sentiment: "It felt like a blessing to practice because we had so much fun." Michael Jordan said, "I have failed over and over again in my life. And that is why I succeed." Researchers have even discovered that people who have high levels of stress without depression are some of the happiest people in the world. They are also the people who are most likely to view their lives as meaningful.

Stress is unavoidable when we chase after a worthwhile dream in alignment with our sense of purpose. It would only be natural for it to be associated with long-term joy and meaning. "The women that I've worked with that medal are the ones that really enjoy the process. They love that idea of pushing the limits and learning and being challenged emotionally and physically," said Olympic trainer and University of California at Berkeley's swim coach Teri McKeever.

What these masters in their respective art forms teach us is that the discomfort of fear, failure, and risk is simply a matter of the frame we have set for them. Their stories, like all the others in this book, are references for your remembering self to use, priming your subconscious mind to embrace struggle in any form. Implanting an empowering belief about suffering will make it easier for you to choose it again and again. It will allow the remembering self to tell a different story about the time you fell off your bike so you can get back on it again.

"Struggle is not optional—it's neurologically required. In order to get your skill circuit to fire optimally, you must, by definition, fire the circuit sub-optimally; you must make mistakes and pay attention to those mistakes; you must slowly teach your circuit," writes Coyle. We all actually know this, and we learned it at quite a young age.

Researchers from Norway and America conducted a study to determine what factors most contributed to a baby's ability to improve at walking. They looked at every possible criteria, including height, weight, age, genetic composition, and brain development. None of these proved to be the defining elements for success. What mattered most was the amount of time the babies spent attempting to walk.

Reading about Fearvana can only take you so far. The real value is in the doing. We live in a world constantly trying to make our lives easier, but if you want success and meaning in your life, don't avoid struggle. Welcome it. Seek it out every day. It can be as simple as eating or brushing your teeth with the opposite hand. These small struggles train your mind to get comfortable with the uncomfortable, which will help you ingrain the belief challenge is fun. They also force you to become more conscious in your actions, training your human brain to become stronger. This mindset of embracing suffering is the doorway into Fearvana, but it is equally essential to survive what lies waiting for you beyond that door.

The Burden of Triumph

"Victory is always possible for the person who refuses to stop fighting."
—Napoleon Hill

Ever since I conquered my addiction to drugs to do something more meaningful, the real challenge became handling life after my Fearvana moments. The inability to manage the transition back to "normal" life after coming home from the war in Iraq, climbing in the Himalayas, or crossing the ice cap in Greenland led me to the darkest of places.

The struggle for Fearvana is not just about reaching it, but figuring out how to continue reaching it once the experience is over. Doing that requires "an evolved capacity to experience anticlimax faster, sharper, and deeper than the rest of us," says Matthew Syed, author of *Bounce*. "Anticlimax is something we have all experienced, but it is striking just how quickly top performers come down to earth after winning a major title; remarkable how rapidly they emotionally disengage from a goal they may have spent years striving for."

It is not just failure top performers interpret differently from others; it is victory as well. The best in any field seem to relish the joy of victory for only a few moments before moving on in their quest for the next one. Minutes after winning his ninth premier league trophy with Manchester United, Sir Alex Ferguson said, "I'm already looking forward to next season. Let's get on with it.

I'm looking forward to going on to win a European trophy as well as pushing for the league." The following season, Manchester United once again won the premier league and the top title in Europe.

Ironically, to become a top performer and stay one, you not only have to struggle through years of training, but you also have to be able to quickly forget the victory you worked so hard to achieve. If our triumphs produced too much fulfillment, it would destroy the internal drive that keeps us striving toward our next goal. Contentment and satisfaction are enemies of mastery. They are pathways to mediocrity.

"A number of people I've been in touch with following the Olympics, people who'd succeeded, said the same. They felt quite depressed, almost like a sense of loss," said peak performance mind coach Steve Peters. The low moments after winning push us to seek out the next high and reach greater levels of success. They set us up to want to experience Fearvana once again. Without that low, Michael Jordan would have been happy with just one NBA ring instead of the six he won by the end of his career.

When Jack Canfield's book *Chicken Soup for the Soul* became a worldwide phenomenon, he did not rest on the glory of that success. He continued stepping outside his comfort zone and has now published more than one hundred books. Collectively, his books have sold over 123 million copies in North America alone. Mastery in any endeavor requires letting go of one victory to pursue the next one. This letting go must happen on a moment-to-moment basis, every day, every week, every month.

During my ultramarathon training, I can't stop for long to celebrate hitting my mileage goal for the week. If I want to be the best version of myself, when Monday morning rolls around, I have to let go of last week's victory and start again. By releasing that old self, I get to create a new self that is reborn through Fearvana.

In the wise words of Bobby Maximus, "Satisfaction is the end of the line. It will kill any desire you have left within you. Once you become satisfied, you'll stop improving, and if you're not growing, you're dying. Words and thoughts like 'I'm good enough,' 'I've done enough,' 'I'm successful enough' will lead you straight to the land of the mediocre. Never let your fight end. Always strive to do more and to be more. Never rest. Never be satisfied." Continued success requires repeated entries into Fearvana. You don't get there once and find yourself on the right path for the rest of your life. The beauty is that within each entry lies our next awakening.

The Doorway to Enlightenment

"Out of suffering have emerged the strongest souls; the most massive characters are seared with scars."

—Khalil Gibran

To keep moving forward on that slow, strenuous journey to unending growth, it is essential to learn how to balance what psychologist Philip Zimbardo calls the two "time perspectives": the present and the future.

Present-oriented people live only for the now, easily succumbing to instant gratification and moment-to-moment pleasures. As a result, "presents" are more prone to addiction and mental health problems. On the other hand, for future-oriented people, "tomorrow's anticipated gains and losses fuel today's decisions and actions," writes Zimbardo. Inevitably, they will be more willing to suffer now and reap the rewards later. But an entirely future-oriented life is one that forgets to soak in those three-second chunks that make the experiencing self happy, which is why, at some point, "futures" often run out of steam and break down.

"After '08, mentally, I was over. I didn't want to do it anymore," said Michael Phelps after winning eight gold medals that Olympics, "but I also knew I couldn't stop. So I forced myself to do something that I really didn't want to do, which was continue swimming." During an interview with Phelps for *USA Today*, reporter Nancy Armour described his future-oriented lifestyle: "For most of his life, Michael Phelps chased the next thing. The next meet. The next race. The next record. The next gold medal."

Eventually, Phelps' burnout led to getting arrested twice for drunk driving. After the second arrest, he checked himself into a six-week rehabilitation program where he began delving into the third time perspective essential for living a full life: the past. "I wound up uncovering a lot of things about myself that I probably knew, but I didn't want to approach," Phelps said about his time in rehabilitation. "One of them was that for a long time, I saw myself as the athlete that I was, but not as a human being."

The remembering self is a key player in shaping the decisions we make, so we need to know how to tell stories of our past that allow us to thrive in the future. These stories are often buried in our implicit memory. For Phelps, one such story included feelings of abandonment by his father. During his therapy, Phelps uncovered all these subconscious obstacles that laid dormant within him, buried by alcohol and the rigors of training.

Gay Hendricks, bestselling author of *The Big Leap,* calls these subconscious obstacles "the Upper Limit Problem." He says solving this one major problem in our lives "is possible only by a leap of consciousness. You dissolve it by shining a laser like beam of awareness on its underpinnings—the false foundations that hold the Upper Limit Problem in place."

The more we become conscious of our buried stories and bring them into the explicit through self-awareness, the less likely those stories will hold us back. "I don't feel like I'm carrying weights around anymore. Whatever I'd been holding inside of me, I was able to get it out and was able to start fresh," said Phelps after exploring his subconscious in therapy. "For the first time, I was able to look back at my career and be really excited and be proud of what I did. Because I never had the chance to do that before." Phelps has won more Olympic medals than any other athlete in history. He worked hard, pushed his limits, repeatedly entered into Fearvana, and continued achieving unparalleled measures of success, yet Phelps was plagued by forces that kept him unfulfilled.

Throughout this book, we have discussed the value of fear and how to harness it to become successful. However, as we have just learned from Phelps and the anticlimactic nature of victory, in the long run, that is not enough to find real inner peace. Without awareness as to the forces driving our actions, even a worthy struggle can be a form of avoidance. For me, it became a strange kind of comfort zone that, until recently, felt near impossible to escape from.

During one of my trips to India, I learned an uncle of mine needed a new liver, or he would die. Four people volunteered to donate their liver to him. I was one of them. We each had to undergo a series of tests to determine if we could be viable donors. Of the four people, two passed: his wife and me. During the consultations with the doctors, they informed us that removing the liver would require removing the gall bladder as well. This presented a significant obstacle. Through extensive research, my family and I discovered life without a gall bladder could cause major digestive problems. I couldn't afford to have trouble with digestion while running an ultramarathon, climbing a mountain, or skiing across an ice cap. That lifestyle had been my salvation on more than one occasion. It saved me from the brink of my own death. If I had to give it up, what would that do to me?

In light of these concerns and knowing his wife passed the initial tests too, I backed off as a volunteer. I told myself, since we both were viable donors, the burden must fall on her as his spouse. But I know in my soul that if for any reason before the surgery she could no longer donate her liver, I would have still given him mine. I could not have let him die.

Ultimately his wife donated her liver and today they are both doing well, but the experience revealed a great deal about me. By spending time with myself and processing my decisions, I realized that deep down a large part of me wanted to be the only viable donor. I wanted to be the hero. I wanted that scar on my chest as a testament to my warrior spirit. I hadn't been shot on the battlefield. I needed another war and the wounds that came with it to earn my place on this planet.

Losing my gall bladder would only have meant a new and beautiful struggle to take on. A part of me didn't care what everyone told me. To me, it was a tangible battle I knew I could win. I had done it before when doctors told me I couldn't survive boot camp. I relished the idea of being back in the fight. The challenge of sitting in front of a computer to write a book was not as alluring as donating a liver, losing a gall bladder, and fighting my way back to becoming a peak performing athlete. That would have been a worthy story to take with me to my grave.

As you might guess, this was not a healthy way to approach life. For a long time, I lived with this inside of me. I needed a war to be happy. I thrived in a state of struggle that meant walking on the edge of life and death, but I could not handle the lows between those highs.

The struggle associated with Fearvana is necessary, but it is the doorway to enlightenment, not enlightenment itself. I have known people who struggle for hours in the gym to look good, but deep down, it didn't make them any happier. Struggle can be a tempting way to avoid the inner work that takes place beyond it. It's the twenty-two hours outside the gym where the battle is won or lost. The value of doing that work is it helps us find more meaning and joy in the struggles we choose to pursue, even if those struggles are no different than the ones we once chose as a means of running away from ourselves.

"I'm just more engaged in everything I do in my life . . . I'm back to being the little kid who once said anything is possible," said Phelps about his preparation for the 2016 Olympic games in Rio De Janeiro. "You're going to see a different me than you saw in any of the other Olympics." Phelps won five golds and one silver in that Olympics.

For me, I still climb mountains and run ultras, but I do it from a place of higher awareness. These endeavors are no longer a means of escape to mask my problems. I now push myself to the edge to continue developing the strength to address those problems, whatever form they may take. Today, my chosen pursuits have become a part of my greater self, not the defining elements of it. I know now life is about much more than just those two hours in the gym. I have learned to let go of that feeling of never being good enough by aligning my mind, body, heart, and spirit as one. Who I am today is more than enough, yet there will always be more room to grow. That is the beauty of life. It's limitless.

The back roads in our minds that lie waiting to be explored are unique for each of us. The cycle of action combined with awareness is our roadmap in that glorious journey to an awakened self. Staying in the cycle is how we make repeated entries into Fearvana, despite the lows or the highs. Without engaging our consciousness to choose and reflect on our entries into Fearvana, those moments might well be wasted when we look back on our lives to find the meaning in our stories.

Ultimately, to master the art of suffering to achieve inner and outer success, we must proactively engage the three time perspectives of past, present, and future. When we integrate all of them within the struggle inherent to the

Fearvana lifestyle, we align all parts of our identity into one blissful, unstoppable, awakened self.

Training Exercises

"If you want to achieve your goals you have to do more than just show up. You have to work, you have to grind, you have to suffer."

—Bobby Maximus

We don't always get to choose how we suffer. Life often makes that choice for us. By no means is it easy to smile in the face of suffering we don't actively seek out, such as genocide, war, rape, or the death of a loved one. But even in the case of a traumatic event, no matter how horrific, there are immense opportunities for growth.

Our focus here is about consciously seeking out a worthy struggle in service of a desired goal, so going deep into the nature of trauma goes beyond the scope of this book. However, if you have suffered through a traumatic event, I don't want to leave you hanging. I know how challenging it can be, especially since there is not nearly enough useful information available on how to turn trauma into growth. As such, I have included a bonus chapter that will reframe your entire perception of trauma and teach you how to find the value in it. That chapter is available to you for free at www.fearvana.com/resources. If you have ever come up against any kind of traumatic event, I highly recommend you download the chapter. It will forever change the way you think about the meaning of trauma. For now, here are three training exercises to help you seek out and endure the suffering that will accompany you on your road to greatness.

1. The Suffering Smile

On the journey to creating the life you want, you will face fear, suffering, stress, misery, and pain. To embrace these challenges, it is vital to learn how to make them fun through the suffering smile or, as it is more commonly called, play. According to Dr. Stuart Brown, author of the book *Play: How It Shapes the Brain, Opens the Imagination, and Invigorates the Soul,* play is defined as "an absorbing, apparently purposeless activity that provides enjoyment and a suspension of self-consciousness and sense of time." Sounds a lot like Fearvana, doesn't it?

Fearvana is the intersection of play and suffering. It is when pain and pleasure, the two driving forces of human behavior, unite to create a sum greater than its parts. It is not always easy to find pleasure in pain, but "making all of life an act of play occurs when we recognize and accept that there may be some discomfort in play, and that every experience has both pleasure and pain," writes Dr. Brown.

To make life playful, bring lightness, joy, and fun into every experience, especially the challenging ones. I call this the suffering smile. How do we do this? It's not complicated; we simply choose to. "Play is a state of mind—it's a way to approach the world. Whether your world is a frightening prison or a loving playground is entirely up to you," writes Charlie Hoehn, author of *Play It Away*.

The mindset of play is how your experiencing self gets the immediate return it desires. It also gives you the means to endure the struggle chosen by your remembering self for the delayed return rewards. Play doesn't just make that struggle more enjoyable, it actually improves your brain as well. "Play promotes the creation of new connections that didn't exist before, new connections between neurons and between disparate brain centers," writes Dr. Brown.

Sports and exercise are perfect representations of play because they involve struggle in a game-like setting. They combine play and suffering in the present to pursue a delayed victory. That is why so many of the examples I have shared throughout this book involve athletics. But play can apply to any aspect of life because play itself is not an activity; it is a mindset that will help you before and after Fearvana. Play is how you enjoy those moments between each experience of Fearvana and reenergize yourself for reentry. You need time off from that war against yourself. Rest gives you the means to come back into the fight stronger and harder than before.

When I work with teens, I always tell them not to cram until the moment of their exam. I encourage them to relax and have some fun. "I'm finally relaxing after so long . . . never done this before," one teenager told me after taking much-needed time off. She was a hard worker, but that was part of the problem. She spent so much time and energy stressing out before every exam, she rarely took a break. Her grades suffered as a result. Overexerting the brain will wear it out, as does overexerting the body.

Just as Fearvana gives you courage in other areas of your life, having fun and playing when you are at rest will give you the ability to bring that joy into Fearvana. All you have to do to make play work for you is choose to adopt this mindset of the suffering smile throughout every moment of your life. Don't overthink this and make it more complicated than it is. Choose to approach life as a series of experiments or a game, and it will become just that. The next time you face a challenge, smile and ask yourself, "What is fun about this? How can I make this enjoyable?" You will be surprised by the answers that come to you.

2. Visualize the Pain

A study conducted at UCLA asked three groups of students to choose a solvable problem in their lives that was "stressing them out." The control group was told to think about the issue, study it, and figure out action steps to take to address it. The "event-simulation" group was asked to visualize the problem, picturing all the steps in detail that created it, the environment around it, and what steps they had already taken to deal with it. Finally, the "outcome-simulation" group was asked to imagine themselves on the other side of the problem, feeling the joy, relief, and satisfaction of having overcome it.

The two simulation groups were asked to follow their visualization instructions for five minutes every day. After one week, the researchers found, out of the three groups, those in the "event-simulation" group were the ones most likely to have taken action and sought out the necessary help to solve their problems. By visualizing the event itself, they were also more successful than the other groups at uncovering the lessons and the growth available to them within their problems. Visualizing the process of struggle, as opposed to the outcome on the other side of it, better prepares you to overcome that struggle.

Before a long run, I always imagine myself hitting a low point when the pain will inevitably hit. I then picture myself fighting through the pain to continue running. This has been far more effective to my success than visualizing the joy of finishing the run.

I did the same thing when I quit drinking. Instead of visualizing myself sober and happy, I imagined myself in all the situations I knew would trigger a desire to drink. I then mentally practiced what I would do if those cravings showed up and how I would find joy in those environments without a drink. I did this on the day after I quit drinking, and the memory of that visualization has stayed with me forever. Since then, I have only felt a desire to drink once, and it lasted less than ten seconds. Other than those few seconds, I have never again experienced a craving for alcohol, despite the fact it had been a part of my life for seventeen years.

You can do this no matter what struggle you are facing. It could be sitting down to write, negotiating a raise with your boss, talking to a potential client—anything. Go into as much detail as needed when you imagine yourself in the struggle. Slow it down as well, so you give yourself the time to navigate the problem in your head. Initially, I recommend going into all the details you can think of as slowly as possible. That will better prepare you to face all the possible pain points waiting for you. It can't hurt having more details, so have fun with this and go wild. By visualizing yourself being with the pain, you will greatly improve your odds of fighting through it.

"You stand a much better chance of mentally withstanding war if you can visualize it and prepare your brain for it than if you've never thought of it, never been able to picture it," said Navy Seal and director of mentorship for Naval Special Warfare, Lu Lastra. Visualizing the pain doesn't just prepare you for struggle, it also improves your ability at the task you have chosen as your path of Fearvana.

A group of basketball players showed a 23 percent improvement in scoring free throws simply by visualizing the action of shooting a basketball into a hoop for thirty days. The group that physically shot the free throws improved by 24 percent. Without even touching a ball, the visualization group showed almost the same level of improvement. The brain cannot tell the difference between an action that is vividly imagined and one that is physically taken. When we visualize ourselves shooting a basketball, the same areas of the brain light up as if we were actually shooting one. Imagine the possibilities if you combine visualizing the pain with the act of engaging it.

3. Control Time

The ability to endure suffering is dependent on the level of control we have over how we manage the three time perspectives. So instead of letting time control the outcome of our lives, let's take control of time. In chapter 5 on memory, I taught you a technique to time travel and change your past. Now, as we enter the awakening portion of our journey, keep using that technique to better your future. That exercise will help you seize control over the conscious and subconscious effect of your past.

Along with the suffering smile, take control of your present by reducing it to the single step in front of you. Break down the entire journey from where you are now to where you want to be into smaller chunks. Think about what is the smallest possible step you can take to get to your goal. Then focus your energy, attention, and consciousness on the next immediate chunk. As you progress through each step, celebrate and acknowledge yourself for your hard work, not just your results.

A study by Dr. Ayelet Fishbach and Dr. Jinhee Choi from the University of Chicago found students who focused on the process of working out, as opposed to an end goal, like losing weight, not only had better results in the gym, but they enjoyed the experience of working out more as well. Focusing on the process over the outcome doesn't just apply to visualization, it applies to taking action as well. Remember the concept of target shooting?

The last chapter helped you take control of your future self by getting clarity on your desired future. Healing your past, chunking your time, and visualizing the pain will help you make that future a reality.

Finally, leverage the assets available to you in your human and animal brain to take control of all the three time perspectives. To make the past, present, and future guide you on your path of Fearvana, keep practicing being aware of your emotions and proactively engaging them at will. Emotions play an essential role in memory and learning, so you want to be in as much control as possible of how and when they are being recruited.

If you want to lose weight, but hate going to the gym, proactively practice smiling and focus your attention on finding joy in the moment every time you work out. Smile at the thought of the person you will become as a result of that session. Anchor these emotions to all the actions you want to make consistent. Simultaneously, anchor emotions of pain, sadness, guilt, or regret to the past self you no longer want to be and leverage those emotions to drive you forward. Ultimately the objective is to combine the two core emotional forces of pain and pleasure to make them both work for you in the past, present, and future.

Chapter 11

THE MOST IMPORTANT
HABIT OF ALL

"Character cannot be developed in ease and quiet. Only through experience of trial and suffering can the soul be strengthened, ambition inspired, and success achieved."

—Helen Keller

In 1980, two years after becoming the first person to summit Everest without supplementary oxygen, Reinhold Messner repeated the feat while alone on the mountain. "When I rest, I feel utterly lifeless except that my throat burns when I draw breath . . . I can scarcely go on. No despair, no happiness, no anxiety. I have not lost the mastery of my feelings, there are actually no more feelings. I consist only of will," Messner wrote about that solo ascent. "After each few meters this too fizzles out in unending tiredness. Then I think nothing. I let myself fall, just lie there. For an indefinite time, I remain completely irresolute. Then I make a few steps again." Through an act of pure will, Messner fought against the specter of death that loomed all around him—and won.

Throughout this book, I have shared many stories with you about people like Messner. I have used these stories because, in my own experience, nature is the ultimate playground in our quest for self-discovery. It is pure. It cares not for our comfort or well-being. It plays no games and has no hidden agendas. It simply is. The indifference, beauty, and hostility of being in the natural world creates an environment ideal for Fearvana.

Secondly, I have shared these tales of high-risk adventure because they lead us to the edge of life and death. When the consequences for failure mean extinction, it demands we bring out the best in ourselves. The desire to live beyond the present moment inspires the kind of extraordinary effort that reveals our limitless capacity for greatness. Such experiences force an expansion of the will that constructs a new, far more powerful self-image to bring back into "normal" life. This is not to say you have to risk your life to strengthen your will. As long as an experience pushes you to your breaking point and beyond, you can grow from it.

Enter the Will

"The difference between a successful person and others is not a lack of strength, not a lack of knowledge, but rather a lack of will."

—Vince Lombardi

Willpower is the act of taking control over consciousness, or what is more often referred to as self-control. It is the capacity to take action in service of your growth when you don't feel like it. After years of research, Dr. Roy Baumeister found that "most major problems, personal and social, center on failure of self-control: compulsive spending and borrowing, impulsive violence, underachievement in school, procrastination at work, alcohol and drug abuse, unhealthy diet, lack of exercise, chronic anxiety, explosive anger." Ultimately, he concluded, "Self-control is a vital strength and key to success in life."

Self-control is necessary in our quest for growth and happiness, but only a rare few master this ability. In a survey with over one million people from around the world, researchers discovered self-control was rated as the lowest of their personal strengths out of a list of twenty-four character traits. Similarly, a lack of self-control was rated as the highest of their faults. This disease of the mind is called *akrasia*, which is defined as a "state of mind in which someone acts against their better judgment through weakness of will." Akrasia is why we procrastinate on everything from doing homework, to working out, to writing a book.

Seizing control over our consciousness to take action in service of our long-term, consciously generated goals is not easy. The human brain is slow precisely because it takes hard work and energy to put it into action. That is why we live most of our lives on autopilot. "A general law of least effort applies to cognitive as well as physical exertion," says Dr. Daniel Kahneman. "The law asserts that if there are several ways of achieving the same goal, people will eventually gravitate to the least demanding course of action. Laziness is built deep into our nature."

Remember, free will is limited. Subconscious forces, such as our habitual patterns and the beliefs we hold onto, control almost the entirety of our life. This is not a bad thing when we learn how to direct those forces toward our goals. We want our animal brain to run the show so we can go about our day with speed and efficiency, but we need it to be on the same page as our consciously chosen path. Seizing control over our consciousness is how we mold our animal brain, bringing it into alignment with the person we want to be. The quantum Zeno effect rewires the brain when attention is directed to the changes we want to make, or in other words, when we seize control over our consciousness. For example, by consciously choosing to read this book, you are creating new, empowering beliefs in your subconscious about fear, pain and suffering.

But the war is never won. Even when we mold our animal brain to direct us onto our chosen path, upon reaching the next stage of our evolution we will once again find ourselves outside our comfort zone. The animal brain will not be programmed to navigate that territory, and it will retreat to its natural state of laziness, unless we consciously fight back. Each new awakening we seek requires the reshaping of our subconscious through the activation of our consciousness. When I decided to run across every country in the world,

I had to implant new beliefs into my subconscious about the possibility of accomplishing something so audacious. Those beliefs ensured my subconscious was on board with the program, allowing me to endure the intense suffering of training for such an endeavor.

Mastery demands a relentless devotion to the constant battles against our natural state of laziness. That part of us wants nothing but a steady stream of instant gratification. The experiencing self wants joy in the moment; it doesn't care about the happiness of the future self.

We make plans to lose twenty-five pounds or write a book or become a CEO of a Fortune 500 company because of the stories the remembering self tells us, based on the references it has gathered from our own lives and the lives of others. The remembering self is the one that makes decisions for our future self. But when it comes down to taking action, the law of least effort directs the experiencing self to step in and decide relaxing in front of the TV would be much more enjoyable than going to the gym or sitting down in front of a computer or completing a resume. That is, unless we do three things:

1. We train our experiencing self to see suffering as blissful, which, as we learned in the last chapter, is accomplished through Fearvana.
2. Create an environment that allows us to conserve our willpower for when we need it most.
3. Destroy akrasia by strengthening the will. Self-control allows the remembering self to overpower the experiencing self's desire for pleasure and make the decision to suffer now, reap the rewards later.

In what has now become a famous experiment, a group of researchers from Stanford conducted a long-term study with hundreds of children, starting when they were four or five years old. One at a time, a researcher brought each child into a room and placed a marshmallow in front of her. He then told the child he would leave the room and if she did not eat that marshmallow, she would be rewarded with a second one as soon as he got back. However, if she ate that marshmallow, she would lose out on the second one.

Once the researcher left the room, some children couldn't wait and scarfed down the first marshmallow. Others squirmed in their chairs, trying to resist the temptation, but eventually gave up before the researcher returned fifteen minutes later. Yet, a few others found the strength to resist the marshmallow and hold out for that second one. The researchers then followed this group of children for more than forty years and found the children who harnessed their willpower to wait for the second marshmallow found success, no matter how the researchers chose to define the term.

When both the experiencing self and remembering self become accustomed to suffering, they no longer resist making decisions that involve hardship. A positive relationship to physical and psychological suffering is essential to strengthening the will. The problem, however, is that our pool of willpower is limited, and we waste it all the time without being consciously aware we are doing so. A follow-up study by Baumeister found "people spend about a quarter of their waking hours resisting desires." Imagine spending six hours a day doing squats. You would collapse.

Resisting desires wastes our cognitive energy by forcing us to constantly make decisions. Think about your day, where are little decision battles depleting you? Should you jump out of bed or hit snooze? Should you meditate and exercise in the morning or skip it to watch TV? Should you work on your writing project or go onto Facebook? Should you eat dessert or not? Spend a day tracking every time you make a decision, and you will notice you fight *a lot* of cognitively exhausting decision battles.

This is known as decision fatigue, or willpower depletion. To test its effects, Baumeister and his team put a group of starving college students in a room filled with the irresistible scent of freshly baked chocolate chip cookies. The researchers then placed a bowl of chocolates, cookies, and radishes in front of the students. One group of students was told they could eat whatever they wanted. The other was told they could only eat the radishes. The latter group eyed the cookies in agony. Some even picked them up to savor their aroma, but they could not take a bite.

This made it evidently clear that these students wanted to eat the cookies and needed to exercise self-control in order not to. Both groups of students were then taken to another room to solve a puzzle they did not know was unsolvable. The radish-eating group gave up after only eight minutes. The group of students who were allowed to eat the cookies gave up after twenty minutes. A control group of hungry students that was offered no food at all also lasted twenty minutes. This demonstrated that the only reason the radish eaters lasted less than half the time as the other two groups was because they had been forced to drain their willpower.

A number of other experiments have further proved that when we use willpower to resist thoughts, actions, and emotions, or even when we push them to their limits, this reduces our ability to employ our willpower at a later time. No matter the activity, using willpower in one setting drains it in another. We only have one pool that supplies all our willpower. Like any other muscle, it can be exhausted. But, like any muscle, it can also be strengthened.

In a follow up to the marshmallow experiment, researchers from the University of Rochester replicated the exact study with one difference. Before giving the children the first marshmallow, the researchers broke them up into two groups. Group 1 was placed into a reliable environment, meaning the researchers promised them things like the best set of crayons and stickers and then delivered on their promise. Group 2 was placed into an unreliable environment. The researchers gave those children a set of crayons and stickers, then promised to bring them better versions of both but never did.

How do you think this affected them when they were given the marshmallow test? Group 2 believed the world was unreliable. They had no reason to trust that the researcher would bring them a second marshmallow. On the other hand, group 1 was taught that the world is reliable. In their mind, if they waited for the second marshmallow, they believed they would get it. This belief led to them waiting four times longer for the second marshmallow than group 1. This not only taught them that exercising self-control is worthwhile, it also taught them they had the strength to do so.

In both groups, their beliefs were implanted into them by their environment. One experience lasting no longer than a few minutes altered their abilities to exercise the most important skill for success: self-control. Once again, we see the shocking and scary effects the environment can have on us. "One of the central tenets of this book is that our behavior is shaped, both positively and negatively, by our environment—and that a keen appreciation of our environment can dramatically lift not only our motivation, ability, and understanding of the change process, but also our confidence that we can actually do it," writes Goldsmith in his book *Triggers.*

Never forget that your beliefs, your thoughts, and your emotions are not inherently "real." Your environment inserts them into your subconscious. You then make them real, but you can also undo that effect. As the growth mindset and the principles of neuroplasticity illustrate, everything in your mind is malleable. No matter where you are now, you can change your beliefs to strengthen your willpower and form a positive relationship with suffering. You do have the power to ingrain the mindset that pain now equals pleasure later, instead of giving in to the hedonistic experiencing self.

The Limitlessness of the Human Spirit

"Courage is not simply one of the virtues, but the form of every virtue at the testing point, which means at the point of highest reality."

—**C.S. Lewis**

To thrive in the space between each entry into Fearvana and build up the boldness to keep reentering it, it is essential to ingrain a habit I call cultivating courage. Ironically, cultivating courage is not actually a habit, yet it is the foundation from which you build all other habits. Charles Duhigg, author of *The Power of Habit*, calls such habits "keystone habits." They "are more important than others because they have the power to start a chain reaction, shifting other patterns as they move through our lives. Keystone habits influence how we work, eat, play, live, spend, and communicate," he writes. They "start a process that, over time, transforms everything."

No matter what habit you are trying to develop, it always begins with an act of consciousness. You might place your dental floss next to your toothbrush as a strategy to make flossing more of an automatic behavior, but the act of making that choice and moving your dental floss was a conscious one. You might remove all desserts from your fridge so you don't give yourself the option of eating unhealthy food, but, once again, that initial decision was an act of consciousness.

Exercising the will is the first step to making any behavior automatic. Cultivating courage is about making that act of exercising your will a normal part of your life. The secret to beating your lazy brain is to make a habit of choosing suffering, choosing to be uncomfortable, and choosing to engage your fears. In my research, I have rarely, if ever, seen anyone address the habit of cultivating courage. I believe this is primarily because the habit itself is somewhat counterintuitive. Cultivating courage is an act of engaging our human brain to exercise self-control, which is inherently not an automatic process and, therefore, not a habit. It requires the activation of attention in a conscious act of the will, but that is precisely the point. Cultivating courage is the habit that allows us to stop living as machines on autopilot.

"These ideas are stretching the formal definition of a habit: which involves the same behavior or thought in the same situation. For happy habits, we need slightly different behaviors in slightly different circumstances. We need the habit to rise above itself," says Dr. Jeremy Dean in his book *Making Habits, Breaking Habits*. "The ideal situation is an automatic initiation of the behavior, but then a mindful, continuously varying way of carrying it out. A new type of hybrid habit: a mindful habit." Dr. Dean discusses this kind of habit in reference to happiness. We will come back to that later in this chapter. In this context, what it means is that we want to make the initial act of venturing out of our comfort zone habitual. Inevitably, though, once we find ourselves in the unknown, we will need to keep figuring out how to navigate that terrain. Athletes are the perfect example of people who practice this habit.

Basketball player Bill Russell had to rise above anxiety to the point of throwing up on a regular basis. This required a high level of self-awareness, to step outside of the uncomfortable feeling and choose from a place beyond it. Even Ronda Rousey has said she fought above herself. She and Russell both made the conscious act of engaging their will a consistent habit. Like any other athlete, they then had to face a new challenge every time they stepped onto their respective battlefields. This is why athletes are celebrated as metaphors for human excellence and sometimes even worshiped as gods. Their repeated battles with fear and the courage they exert to face it transforms them into superstars in our society. Mark Twight brilliantly describes the paradoxical nature of habituating the will:

> A strong will grows from suffering successfully and being rewarded for it. Does a strong will come from years of multi-hour training runs, or do those runs result from a dominating will? There is no right answer because will and action feed one another. Suffering provides the opportunity to exercise will and to develop grit. Relish the challenge of overcoming difficulties that would crush ordinary men (and women).

Another reason for the lack of conversation on habituating willpower is because of the nature of willpower depletion and the state of our current world. Our environment today constantly requires us to exercise our willpower muscle. "In every shopping entity, in every advertisement, all of them want something from you. And what they want from you is not your long-term well-being," says psychologist Dan Ariely. "What they want is your time, money, and attention right now. It's getting harder and harder because our apps are on the phone, websites are popping, so this question of resisting temptation is something that is becoming more and more difficult."

Although the exact number varies by source, it is clear we are exposed to a significant number of advertisements every day. The most accurate estimate seems to be about five hundred to seven hundred a day. That is a lot of temptation we have to contend with. Many of us become victims to this onslaught of information. We saw this occur in the study on anchoring where the number of tomato soup cans mentioned on a sign influenced the decision on how many cans to buy.

Or take the example of when actress Blake Lively carried a handbag on the show *Gossip Girl*. Within a month of airing, sales of that brand shot up by 30 percent. Do you think people consciously chose to buy that handbag because they really wanted or needed it? Or do you think that because of the high value our culture places on celebrities, their brains made a subconscious, emotional decision to make the purchase?

We constantly make such decisions based on the effect that external information has on our animal brain. We might be working as hard as we can to lose weight, but an ad for McDonald's comes on, and suddenly we find ourselves craving a Big Mac. Now we have to either exercise our willpower and battle that urge or surrender to it.

This is why clarity of purpose and action is essential to mastering life between each entry into Fearvana and finding the strength to make the reentry. When you know exactly what you want out of your life, you can then start to create an environment conducive to that path. Should you choose to immerse yourself entirely into it and let it consume you, that ad for McDonald's won't matter anymore. It will no longer have any effect of you.

Creating a conducive environment for your path of Fearvana will leave no room for distractions from the outside world to invade your consciousness, allowing you to save your pool of willpower for when you need it. You can either become a product of your environment or a creator of it. Clarity gives you the focus to strengthen your will and exercise it for a specific purpose.

In doing so, you might ultimately discover there are no limits to your willpower. Multiple studies have shown that even when pushed to our limits, it is not our muscles that give way, but our minds. In one study, researchers pushed a group of cyclists to the point of complete exhaustion. Even though the cyclists felt like they were going to die, only 30 percent of their muscle fibers were being used. "Fatigue is less an objective event than a subjective emotion," writes Daniel Coyle. Our bodies have the strength to last as long as our spirit does. The real test of will is measured only by the doing. It is explored in the harshest of environments that demand we exceed our limitations just to survive.

Those who live on the very edge of the human potential intimately understand that the point of failure of our willpower can be stretched far beyond any limits we could possibly define or fathom. The human will is inherently unquantifiable. Thankfully so, because no matter how far we progress in life, we never stop needing it.

The Eternal War for Mastery

"It is better to conquer yourself than to win a thousand battles."

—Buddha

In one of her studies, Dr. Sian Beilock asked a group of elite soccer players to focus on and pay attention to the side of the foot that touched the ball as they dribbled through a set of cones. When using their consciousness to navigate through the cones, they made more mistakes and dribbled more slowly than they would have otherwise. However, when she asked a group of beginner soccer players to do the same thing, their performance dramatically improved. The newer you are to a task, the more you need to use your human brain to become better at that task. For a beginner in any endeavor, just showing up to train takes an act of willpower.

When training some of my little cousins who had never exercised before, I couldn't bring them into the gym and make them suffer through the intensity of a Fearvana-style session. They would never want to exercise again. I had to slowly inoculate them into developing a positive relationship to pain. Initially, they just needed to get in the gym and keep doing something small to work their way up to the point of relishing true suffering. Eventually, one of them trained with me enough to start taking on intense interval sessions that made him throw up. He learned to enjoy the feeling of hard work that pushed him to his limits. The further away you are from mastery in any endeavor, the more important volume becomes over energy expended.

As a young child, David Beckham practiced kicking a soccer ball for hours on end at his local park. In describing his early tennis training regimen, Andre Agassi wrote, "My father says that if I hit twenty-five hundred balls each day, I'll hit seventeen thousand five hundred balls each week, and at the end of one year I'll have hit nearly one million balls."

The quantum Zeno effect ensures, as you get better at any activity, you build neural networks that imprint the skill into your animal brain. As you develop mastery, implicit memory takes over, and your actions become habitual, just like when you ride a bike. As Dr. David Eagleman states in his bestselling book

Incognito, "The task becomes burned into the machinery." Getting to that point requires a consistent use of willpower to consciously focus your human brain on the task at hand, as Beilock demonstrated in her study with amateur soccer players. It also requires a consistent use of willpower to step onto the battlefield.

Regardless of how skilled you may be at exercising your will, the habit of cultivating courage is not just about the courage to do the big, scary things, like running a fifty-mile race. It is also about having the courage and self-control to run those shorter three-, five-, and ten-mile training runs, especially when you don't feel like it.

Even when you become an expert, the monotony of regular practice does not stop. Consider Stephen Curry, who was the first NBA player to win MVP by a unanimous vote and is considered one of the greatest basketball players of all time. He shoots the ball over one thousand times during practice every week and two hundred times before the tip-off at every game.

Fearvana is the union of deliberate practice and flow. In the space between those intense Fearvana-style training sessions, we must take the smaller steps of "regular practice" to prepare for the next one. Showing up to build volume can sometimes be boring. Other times, it can feel as challenging as any hard day on the battlefield. Either way, it doesn't matter what it takes or how it feels. All that matters is whether or not you decide to show up. You will never stop having to make that decision.

Momentum can be our best friend or our worst enemy. The experiencing self is easily tempted by distractions, such as alcohol and fast food, for example. When we lose focus, we start creating new stories for our remembering self about what it means to be happy. Our malleable memories adapt to whatever is most useful for our present self. If we put up our feet and take a break for too long, slowly we will find ourselves seeking out immediate pleasure once again. At that point, our remembering self has all but forgotten that pain now equals pleasure later, and it becomes easy to curl up into our comfort zones.

Every single day, you will fight a battle between the right way and the easy way. The cumulative effect of the choices you make in that everlasting war against yourself will send you down the path to an ordinary life or an awe-inspiring one. As Darren Hardy, bestselling author of *The Compound Effect* and the publisher of

Success magazine, said, "Earning success is hard. The process is laborious, tedious, sometimes even boring. Becoming wealthy, influential, and world-class in your field is slow and arduous." Mediocrity or mastery? You get to decide.

Should you choose the path of mastery, the objective in surviving the arduous journey is to activate the human brain to the extent that it molds the animal brain to act in line with our conscious goals and sense of self-identity. After completing the transformation, we want our newly developed animal brain to regain power so it automatically guides us on our chosen path with speed and efficiency, like the way it did for those professional soccer players in Beilock's study.

But once a behavior is automated, we must once again bring the human brain into action to achieve the next level of growth. Long-term, automated habits are necessary, but at some point they lead to a plateau. The consequence of mastery for those that reach it is the "inability to escape the confines of their own habitual ways of thinking," writes Dr. Dean. "Experts are good when new problems follow similar patterns to old ones, but can become blocked when they don't." This is an inevitable cost of developing familiarity with a task to the point that it becomes automated.

As we used to say in the Marines, when you train yourself to become a hammer, everything looks like a nail. With expertise comes the inability to step away from the very patterns set by the years of work it took to create them. As with any subconscious pattern, when we become caught in its grip, we move through life on autopilot. If we don't keep adding variety to our life, automation impedes growth. To avoid this, we must continue to venture out into the unknown, overloading ourselves to take on more than we think we can handle.

For someone like Stephen Curry, routine training has become normal and habitual. At his level of the game, it is simply not enough. To reach new heights, Curry has to enter new territory and explore new frontiers. "We overload our sensory system, nervous system, in our training . . . Once you get to a certain level, all that hard work that you put in resets itself," said Curry. "You have to reestablish yourself. Just because you got to that certain level doesn't mean you made it. You have to work even harder to be successful . . . It's all about finding a different way to improve. It's confidence, it's practice, it's about just pushing yourself to new limits."

Think of this process of mastery like learning math. When you first start learning, you have to think about the answer to 5 times 3. Then you get to a point where you can answer basic problems without thinking. As you start progressing toward more advanced math, like calculus, you have to once again start thinking about it. Constant improvement requires the re-activation of consciousness to once again test and engage the will. Reshaping old patterns not only leads to continuous growth, it makes us happier as well. "Any activity we carry out to try and increase our happiness, if it eventually becomes completely routine, can soon be unconscious and unnoticed," writes Dr. Dean.

My wife has been victim to this phenomenon on multiple occasions. When we visited St. Lucia for our four-year anniversary, she was in awe of all the mountains on the island. I arrogantly pointed out to her, "These aren't real mountains; I've seen better." While I appreciated the beauty of the island, I no longer felt the initial sense of awe and novelty she did. I had roamed in the big mountains of the Himalayas, and I made sure to point that out to her. You can imagine how much she loved my "first-world problem of having seen too much" comments about new places we visited together.

Think about it for yourself. The first time you took a sunset walk at your local park, you might have been in a state of absolute bliss. But do the same thing every day for weeks on end. How does that walk then feel?

Once automated, happiness, like growth, no longer becomes possible. We need novelty to be happy and to grow as human beings. Habituating the experience of Fearvana ensures the perpetuation of novelty. "Psychological training and the conscious reshaping of one's mental habits developed and hardened over years and years is difficult," says Mark Twight. "But at some point we realize that repetition—like all training actions—produces finite results, and what got us here won't get us there. We must remove the blinders. A new, different method or emphasis might unlock future progress." To remove the blinders, we must do two things:

1. Find the problem; fix the problem.
2. Find what's right; do more of it.

The journey to mastery is a combination of these two elements being practiced on the action/awareness cycle. As you walk your chosen path of Fearvana and your zone of Fearvana expands, figure out what is working and repeat those actions. Simultaneously, you will keep finding new problems to fix. That is a good thing. If you no longer have problems, you are no longer growing. Once you habituate the solution to one problem, your next awakening involves finding another one beyond your comfort zone to challenge you again.

In every step of your evolution, your self-image will be reborn. Every victory you achieve, no matter how small, will expand the possibilities available to you for the next one. Tapping into the reserves of your will forges a new self that is better in every way than the self that went to bed the previous night. Who you believe yourself to be now is not who you will believe yourself to be after you train in Fearvana. This new belief will control everything about what you do, who you are, and what you have. As Tony Robbins says, "The strongest force in the human personality is the need to stay consistent with how we define ourselves."

If your self-image is that of a marathoner, you will never run a fifty miler. If your self-image is that of a six-figure earner, you will never break into seven figures. One way or another, you will sabotage your progress to stay within the limits of your self-image. For example, when I set out to run ten miles, I can't imagine doing any more. When I set out to run twenty, it feels impossible to run another ten. But if I leave my house with the intent to run thirty, I finish that thirty. What I am capable of accomplishing is entirely dependent on the limits I set for myself.

That is not to say there is anything wrong with running ten miles, being a marathon runner, or making six figures—as long as you make those choices from a place of awareness. Whatever you choose, your self-image controls your destiny, so don't create one that limits your potential. Remove your blinders, and the possibilities available to you become limitless. There will always be more mountains to climb and valleys to crawl out of in this eternal war against your limitations.

Training Exercises

"He or she who is willing to be the most uncomfortable is not only the bravest but rises the fastest."

—**Brene Brown**

Researcher Raoul Oudejans conducted a study in which he placed police officers in simulated battle conditions to see how they would respond. The police officers who trained against real people shooting at them with colored soap bullets fired back with far greater accuracy and precision than police officers who trained by shooting at cardboard targets. This kind of training is known as stress inoculation. By training in a high-stress situation, they taught their brains to better react to future stressful conditions.

Our goal is the same. We want to train our bodies and minds to thrive in a state of stress, fear, or anxiety so we can harness these forces to keep entering Fearvana.

Cultivate Courage

Cultivating courage is essentially the primary action the entire book encourages; it is how you destroy akrasia and start to implant the belief that exercising self-control is worthwhile. As you continue to grow in courage, it will keep improving your level of confidence in your ability to exercise self-control as well.

Cultivating courage means engaging your will to strengthen it. The specific action steps vary, depending on where you are in your life. No matter where you are, it is where you are meant to be, so there is no right or wrong about the nature of the action. On the one end, it could be something like using your less dominant hand to eat every meal or brush your teeth. At the other end, it could be climbing Mount Everest. The actions range from those two extremes to anything and everything between.

My favorite way to cultivate courage is exercise. It's simple enough for almost anyone to do anywhere, and it has profoundly positive effects in all areas of our lives. When Sir Richard Branson was asked how he gets so much done, his response was "work out."

Dr. John Ratey, author of *Spark: The Revolutionary New Science of Exercise and the Brain* calls exercise "Miracle-Gro for the brain." He says, "Exercise is the single most powerful tool you have to optimize your brain function. The more fit you are, the more resilient your brain becomes and the better it functions both cognitively and psychologically." Exercise is one of the best ways to build yourself psychologically, physically, spiritually, and neurologically. It is the all-in-one solution to building a better brain and a better you. Here is Mark Twight on how to use exercise to build yourself:

> There should be one to two efforts each week that are true tests of will over fitness, i.e. a workout that is daunting enough to require total presence and engagement. This is not the 10 x 1-minute hill repeat session at a pace you *know* you can handle. This work/rest structure and pace should make it impossible to finish all ten intervals. Sometimes you can govern it yourself. Sometimes you need another person . . . It's easy to throw in the towel. Get in the habit of not doing so. I like to set up workouts that I know will produce an internal negotiation. When I develop the habit of winning the internal argument between quitting and not quitting, between backing off and stepping on the gas, between one more interval or one less, then I am training my mind to become an ally and not the opposite.

Setting up such workouts is terrifying. It's meant to be. It allows for not only physical but, more importantly, psychological growth as well. No matter how many times we test our limits, the fear and suffering will not go away. In fact, that is the purpose of these routines—to build the habit of exercising the will.

The other way to cultivate courage is up to you to discover. Find your zone of Fearvana, and venture into it repeatedly. As you do, your zone will keep expanding. It is within that zone where we find the means to succeed. Without entering it, we remain static. If we go too far beyond it, we paralyze ourselves with fear. In both scenarios, we set ourselves up for failure.

Place yourself in situations that scare you just the right amount and see what happens. When the fear strikes, remember, your goal is not to find ways to calm down, relax, or let go of it. Your goal is to use that fear to become the best version of yourself. Your nerves will help you perform better if you decide they will.

Play with these moments to see how your mind currently responds to fear. Practice choosing how you want your fear to serve you, depending on the situation. Do you need more adrenaline, more connection, more meaning, or more of a big-picture outlook? Whatever you need, entries into your zone of Fearvana will teach you how to summon that response at will. There are no limits to what you can do and who you can become. Use your imagination and transform every day into an experiment to cultivate courage.

The first step into fear, and to making any change, demands the engagement of our will. Cultivating courage will help strengthen that will, but remember willpower is an exhaustible resource. We don't want to tap into the willpower pool unless we absolutely need it. As we have learned, to conserve our willpower and prevent decision fatigue, we need to create a reliable environment. That way, when we consciously choose to access the will in pursuit of Fearvana, we have a full supply to work with. Without an environment conducive to your path of Fearvana, you might not even know when and why you are falling off track.

In one study, researchers found that something as simple as eating from a larger tub of popcorn led to people eating an average of 173 more calories than people who ate from a smaller tub, despite both parties being told they could eat as much as they wanted for free. The popcorn was intentionally made to be stale so their decision to eat more had nothing to do with the flavor of the popcorn.

The participants claimed the size of the tub would in no way have affected their decision, yet it clearly did without their awareness. When our lazy, not-so-intelligent elephant brain sees a larger tub, it gets excited and eats more food. We need the smart rider sitting on top of the elephant to hide that tub and control what the elephant sees. So if you want to eat less, you could start by reducing the size of your plates.

If you are trying to lose weight, it is going to take strength and willpower at some point. You don't want to sabotage yourself with an environment counteracting all your effort and forcing you to waste mental energy making decisions about your diet. You want those decisions to be on autopilot. Setting up stringent structures for the mundane tasks in your life will ensure you don't waste any cognitive energy for such minor decisions.

When we conserve our will to use it single-mindedly toward the one, focused effort we choose to take on in the war against ourselves, the willpower pool becomes limitless. As Messner demonstrated in his solo ascent of Everest, when we are put to the test, we are capable of pushing ourselves far beyond any imaginable limits. So let's set up some stringent structures.

Stringent Structures

The way to establish stringent structures is to create checklists for as many things as possible, while planning and tracking everything. There is no one right way to do this. I recommend starting by creating a three-year plan for your life that breaks down into a one year plan, a one-month plan, a one-week plan, and a one-day plan. Use your path of Fearvana to do this.

Once you have clarity of purpose and action, schedule the specific activities into your calendar. You want to limit your choices and restrict your options so you know exactly what to do and when to do it, without having to think. Such preemptive strikes triple the odds of your success.

One way to make this fun is to take on weekly or monthly decision challenges. Millionaire entrepreneur and philanthropist Marie Forleo calls this a decision detox. The way it works is you create rules for yourself that eliminate decision making from certain areas of your life. You could eat the exact same salad every evening for a week or tell yourself no work other than writing for the next week or do one hundred pushups before you get to eat breakfast. These rules will prevent decision fatigue from draining your willpower pool by giving you clear guidelines for actions to take in areas of your life you are struggling with.

Within your overall plan, create checklists for each individual goal. For example, to finish this book, I created a series of checklists for everything that needed to be done throughout the process. Once I completed that checklist, I created a new one for marketing the book.

But don't forget your animal brain is doing its own thing without your awareness, so you want to ensure it is in alignment with your consciously generated goals by seizing control over your environment. Remove the desserts from your fridge, shut down Facebook, cancel your cable, and do whatever you need to do to walk your path with unwavering focus.

Once you have your plan laid out and your environment under control, then start tracking whether or not you are following the plan. You can track every area of your life, from your diet, to your finances, to the number of hours you watch television—everything! "Tracking is my go-to transformation model for everything that ails me," said Darren Hardy, bestselling author of *The Compound Effect*.

I don't want to overwhelm you, so I suggest you start this exercise by tracking just one behavior you would like to improve. Work on one muscle at a time. Keep a journal you use every day. Before going to bed, spend just five to ten minutes tracking what was accomplished that day and what you could do better. Get a clear plan for the next day so you can wake up charged and ready to hit the ground running.

In your journal, include at least six questions you ask yourself every day. I learned the following set of "engaging questions" from Marshall Goldsmith. Questions direct your focus to finding the answer, and as the quantum Zeno effect dictates, attention shapes the brain.

The questions go like this:

On a scale of one to ten, did I do my best to:

- Set clear goals?
- Make progress toward my goals?
- Find meaning?
- Be happy?
- Build positive relationships?
- Be fully engaged?

You then rate yourself for how you did that day. Note that these questions are framed in a style Goldsmith calls active questions. Of course, the opposite of active questions are passive questions, such as, "Did I have a clear goal?" Notice the difference?

Asking the question in an active format forces you to take 100 percent responsibility for your actions by focusing on the *effort* you extended that day, as opposed to finding excuses. If you consistently see a low score, it will drive you to improve the amount of effort you exert the next day. You can ask yourself more than these six questions as well, depending on your individual goals. I use sixteen questions in my planner and tracker. Feel free to use as many as you need to make progress on your path of Fearvana.

Such stringent structures are especially important in facing long-term fears, like building a business, as opposed to short-term fears, like jumping out of a plane. Within these structures, we then include the paradoxical habit of cultivating courage. What that means is I could plan a forty-mile run ahead of time, but actually doing it requires willpower.

If you would like to see exactly how I do this and the apps I use to facilitate my stringent structures, you can download all my planning and tracking tools at www.fearvana.com/resources.

Chapter 12

LOVE, FAITH, AND FEAR

"We can never obtain peace in the outer world until we make peace with ourselves."

—The Dalai Lama

I walked past the construction workers into Charu's bedroom. Only her eyes turned toward me when I opened the door. They echoed the tender smile on her face. Lying in bed, her legs were covered by a white blanket with pink roses. Two pictures of her son hung on the wall behind her. As I approached the bed, her arms remained crossed. A lady sitting beside her got up and arranged a chair for me. Before leaving the room, she reached down and repositioned Charu's head so she could look at me during our conversation.

For the last twenty years, Charu had suffered from muscular dystrophy. Other than her mouth and her eyes, she was unable to move a single muscle in her body. But she felt everything. From a mosquito on her arm to the broken femurs and outlying bones in both her legs. Every nerve transmitted pain signals to her brain.

During our conversation, Charu told me, "The good news is that my bones broke in such a manner that they will heal shortly, and I will be back on my wheelchair." Regardless of what life threw at her, she seemed to maintain an unwavering positive attitude. That was no easy feat considering the amount of pain she had been through.

Five years to the day before the birth of her son, she lost a child during delivery. The doctors told her that because of her condition, she would not live through another childbirth. But "God was kind," she affirmed. Both she and her son survived the second delivery.

Ten years later, the child was afflicted by a debilitating liver condition that almost ended his life. They fought through it, and once again he survived. For the next seven years, she continued raising him like any good mother would, except she could not lift her arms to hug him.

Then on December 1, 2005, a medical error during a routine procedure killed her only child. "Even if it was only for a short while, I wasn't even supposed to have lived a day after my son was born, so I am blessed that I got to live with him for seventeen years," she said. Six years after that, in 2011, she lost the last person she felt close to. Charu's sister-in-law used to take care of her every need. Two months before we met, she had died of cancer.

"Sometimes I wonder, God, what are you doing?" Charu told me. Words escaped me. What could I say to her? How could I even pretend to understand the magnitude of her suffering? I then noticed her gray shirt had a black Om symbol on it. Her red bangles sparkled above her well-trimmed nails. It occurred to me she had to ask someone else to put all of that on for her. Glancing back into her eyes, I asked her, "So how do you maintain your faith?"

"This is just my Karmic destiny," she said. "If I smile through it, perhaps I will move up in my next life."

Fear Never Ends

"Don't ask what the world needs. Ask what makes you come alive, and go do it. Because what the world needs is people who have come alive."

—Howard Thurman

No matter what you do or how hard you try, you can never predict what life has in store for you. It is an uncontrollable force that will break you if you let it. Your ability to find meaning in suffering and turn adversity into an opportunity is the most essential skill for survival, success, and happiness.

I have never met someone who embodies this virtue more than Charu. I don't think I would have the courage to face her suffering the way she does. Her inability to move a muscle below her neck forced her to be fully present with the chaos of her consciousness every moment of every day. She faced the constant battle of questioning God's judgment versus having faith in him (or her—I don't pretend to know). She felt constant fear just from being in her position. Yet, she found a way out of it through faith. I can't even imagine the kind of strength that takes.

To this day, my greatest fear is stillness. I am fortunate that my wife is a master at addressing this particular fear. When we visit a park, she will be content sitting still and being fully present in a place of beauty. I, on the other hand, twitch, fidget, and move around. I would rather run around the park instead. She has become my guide in addressing this fear.

My greatest battle now is finding the stillness of peace without taking on the war against myself. This is the step I must take to reach my next level of awakening. You might find yourself in a similar situation, terrified of something that seems as simple as sitting still and doing nothing. That is perfectly normal. Fear will always be a part of the human condition, no matter how far we progress in life.

Despite having attained incredible success as an actor and comedian, Robin Williams struggled with addiction, depression, and a multitude of other mental challenges to the point that he eventually took his own life. "It's just literally being afraid," Williams once said about his desire to drink alcohol. "And you think, oh, this will ease the fear. And it doesn't." When asked what he was afraid of, he said, "Everything. It's just a general all-round arggghhh. It's fearfulness and anxiety."

We have seen a countless number of other celebrities confront similar challenges to Williams, such as Whitney Houston, Lindsay Lohan, Philip Seymour Hoffman, and Lamar Odom, to name a few. Celebrities are people who seemingly have everything a person could want, yet they struggle with addiction, depression, and even suicide.

I don't presume to know exactly what these celebrities were going through internally. I certainly empathize with all of them. My point is simply to illustrate that even those we deem successful by conventional standards wrestle with the same demons the rest of us do. As Csikszentmihalyi says, "A person who is healthy, rich, strong, and powerful has no greater odds of being in control of his consciousness than one who is sickly, poor, weak, and oppressed."

I state this as a cautionary warning to never let your guard down. You will always need to maintain awareness over yourself and your environment, accept what you can't control, take action over what you can, and exercise your willpower to become the person you want to be. Fear will be your constant companion on this journey. If you don't proactively seek out your fears in service of your growth, you will spend your life running from them.

To face fear, we must become aware of where it comes from. You know now that fear is your brain's first defense mechanism. Its presence can shut down reason and rationality. As a result, fear is a powerful force that can be taken advantage of by external forces. Whether it be advertisers, politicians, major corporations, or the people next to you, anyone can exploit and even escalate your fears to influence your decisions if you are not careful.

As long as you practice the steps I have outlined in this book and use your awareness to understand your fears, you will be able to separate yourself from the ones being used to manipulate you, the ones leading you down a dark path, and the ones that arise from within you in service of your growth. The latter of the three are the fears that will bring you one step closer to discovering who you really are. They will guide you into your next awakening. Mark Twight beautifully describes this relationship with fear:

> Overcoming fear is a key to self-knowledge. One must welcome fear, confront it, and become familiar with it. Consistent confrontation with the unknown familiarizes one with fear. Doing the same thing over and over, increasing stress by steps one is certain to equal, and following the same recipe as everyone else allows one to avoid the unknown, to elude

the conflict he or she so desperately needs. These battles, which evolution requires and modern society rejects, may be orchestrated in the dojo, in the gym, on the field, in the mountains, or all of the above and more. The location doesn't matter, and the activity is irrelevant because the conflict is within. Internal struggles are not won at another's expense. Who loses when we defeat ourselves? Some people live the answer every single day.

Only when we come up against ourselves do we discover who we are. Perhaps the greatest part about taking on these battles over and over again is that it doesn't just make you come alive, it brings us together so we can all come alive.

Love and Fear Are Not Enemies, but the Closest of Allies

"The salvation of man is through love and in love."

—**Victor Frankl**

Navy SEALs are known to be one of the toughest groups of warriors in the world. About seven or eight out of every ten people who volunteer for the SEALs fail to meet their standards. The toughest challenge during SEAL training is known as Hell Week. For five and a half days, would-be SEALs are forced to endure lack of sleep, freezing cold temperatures, brutal training exercises, and generally miserable, uninterrupted psychological and physical suffering.

One common trait unites the survivors of this ordeal: "They had the ability to step outside of their own pain, put aside their own fear, and ask: how can I help the guy next to me?" said Navy SEAL Eric Greitens. "They had more than the 'fist' of courage and physical strength. They also had a heart large enough to think about others."

On a neurological level, compassion for our fellow human beings shows up as the neurotransmitter oxytocin. It has been demonstrated that oxytocin levels increase when we hug, kiss, or have sex. Consequently, it has often been referred to as the love hormone, but it does a lot more than that. Neuroscientist Ron Stoop found that by injecting oxytocin into the amygdala of a rat, it was far less likely to freeze in fear when administered an electric shock. The rat still generated a biological fear response, such as increased heart rate, adrenaline, and cortisol, so oxytocin didn't reduce the fear, but it improved the rat's ability to deal with it.

Oxytocin not only gives us the desire, but also the courage to do things such as run into a burning Humvee to save a friend. It strengthens our relationships by making us want to connect with and help people. This is what made the Navy SEALs think about the men next to them, despite their own suffering. In fact, those very moments of fear, stress, and hardship increased oxytocin levels in their brain, which in turn helped them better respond to the rigors of Hell Week. It is the secret seventh chemical released in our brain during Fearvana.

In a study to determine the effects of oxytocin, psychology professor Jennifer Crocker asked one group of participants to think about how a potential job would better allow them to make a difference in the lives of others and contribute to something greater than themselves. The other group thought only about how the job would improve their own lives. The participants with a service-oriented mindset had lower levels of cortisol, and they performed better during the interview, as reviewed by unbiased, third-party observers.

The more you frame the outcomes you want in your life in the context of serving and bettering the lives of your human family, the greater your ability to respond to all of life's challenges. When you create your path of Fearvana, consider how your goals relate to something greater than yourself. You can do this by asking yourself these questions: What impact do I want to have in the world? How do I want to make it a better place before I die? Who do I want to serve most? I like to call oxytocin the human hormone, because it has the power to bring out the best of humanity. In doing so, perhaps we might be able to create peace across the globe.

At the heart of all conflict is the objectification of another. When we stop seeing others as people, it becomes easy to justify violence, hate, and anger toward them. To bridge that divide, we must learn to see others as we see ourselves, as people with fears, hopes, and dreams, just like us. Facing adversity creates an environment conducive to this kind of empathy. Oxytocin gives us the courage to face our fears, but as the Navy SEALs learned during Hell Week, fear and suffering are also powerful catalysts for the release of oxytocin.

After conducting multiple experiments to determine the effect pain has on group dynamics, psychologist Brock Bastian at the University of New South Wales found that "pain has the capacity to act as social glue, building cooperation within novel social collectives."

Assuming you choose to be a decent person with empathy toward your fellow human beings, simply being exposed to someone else's suffering generates a sense of compassion. As you read the stories of people like Alice, Charu, and Dale, you probably felt empathy or admiration for them. When we watch movies or read books, we connect to the characters' joys and struggles. We see ourselves in those characters. When violent attacks or natural disasters take place, people from different walks of life come together to support the victims. Adversity often breeds heartfelt compassion.

The neurological correlate for this empathy is known as a mirror neuron. Multiple studies have demonstrated that when we watch someone else perform a particular action, the same parts of our brain light up as if we were performing that action ourselves. This is especially true when strong emotions are involved. This is why we cry when we see a sad a movie or laugh when others are laughing, even if we don't always understand the joke. The leading researcher on the subject of mirror neurons, neuroscientist V. S. Ramachandran, calls them "'empathy neurons,' or 'Dalai Lama neurons,' for they are dissolving the barrier between self and others."

Away from the comfort of the mundane, on the fateful frontiers of the human experience where the battles against our limitations are fought, we shed our masks, rip away the facade we present to the outside world, and become intimate with all the separate parts of us that make us whole. Fearvana connects us to who we really are. In doing so, it unites us with our larger human family as well. By uncovering our innermost fears, struggles, hopes, and dreams, our mirror neurons open us up to relate to the same deep emotions within the people around us. We get to see ourselves in others and vice versa. Through the lens of this awakened self, it becomes clear that no matter what our race, religion, or nationality, we all share the same experience of life, only in different settings.

During my fifty-five-mile run across Luxembourg, I hit a low point somewhere in the middle of the country. I stopped to put a wrap around my knee to keep the pain at bay. A local man named George saw me sitting on a rock beside the road. A white man in some small town in Luxembourg then invited a disheveled, scruffy looking brown man into his home for water. As we parted ways, he gave me a big hug, kissed me on my cheek, and wished me all the best. George reminded me why I was on this journey in the first place: to bring out the best in humanity and show the world that is who we are.

Similarly, in Iraq, a leader of a small community told us, "I feel sorry Americans have to pay in blood for Iraqi freedom." Another Iraqi man who spent eight years as prisoner of war in Iran during the Iran-Iraq war told me, "Americans and Iraqis must work together to build a better future for us all." As US Marines with full body armor and rifles in our hands, we still managed to connect with and relate to the people whose country we were asked to temporarily occupy. Separate from what you or I feel about the politics of the war, these moments reflected the power of shared struggle to unite two vastly different, potentially conflicting, groups of people.

I have experienced something similar in Marine Corps boot camp, in other runs around the world, in my travels to over fifty countries, and even in my own neighborhood. Fearvana creates the possibility of shedding the sense of otherness that is at the root of all conflict. But, of course, not everyone responds to adversity with selflessness.

If we choose to adopt the belief that fear is our friend, our body will respond in a manner that harnesses fear to experience Fearvana. If we see fear as our enemy, it will become one. Similarly, oxytocin only works in our favor if we believe in its ability to do so. It can either enhance bonding and trust or lead to emotions like jealousy, gloating, and envy. "The hormone is an overall trigger for social sentiments: when the person's association is positive, oxytocin bolsters pro-social behaviors; when the association is negative, the hormone increases negative sentiments," said Professor Simone Shamay-Tsoory from the University of Haifa in Israel.

Belief is everything. If you believe you are the kind of person who would rather serve other people, as opposed to judge and look down upon them, oxytocin will lead to positive social behaviors. If not, you probably have some idea what the darker side of humanity is capable of.

I trust you are reading this because you are the kind of person dedicated to improving your own life and the lives of others, so let's focus on enhancing the good and ignoring the evil. To make oxytocin work in your favor, put yourself in high-stress situations and practice choosing a tend-and-befriend response. This will train your brain to react to fear and stress with love.

In the military, we lived in a world where our well-being meant nothing compared to the good of the group. From boot camp until the time we set foot on the battlefield, we suffered as one, training ourselves to be willing to sacrifice everything for the Marines next to us.

The "no man left behind" mindset became so much a part of us, that during my training for Iraq, they had to instruct us not to run out in open terrain to save a wounded Marine. Snipers often used such a tactic to lure more of us out into their line of fire. The fact that we had to be trained not to run into open fire shows how instinctual the tend-and-befriend response can become in the right environment. I found that to be a magnificent testament to our collective ability to love, and even die, for our fellow human beings. But we were fortunate in that we had each other. Without the comfort of shared misery with a group, it often seems impossible not to feel isolated. Suffering alone can make us believe we are the only people in the world going through pain.

One of my clients who struggled with alcoholism and depression was shocked to learn I had experienced both. Just by sharing my story with him, he gained confidence in his ability to defeat his demons. Before that, it felt like he was the only one struggling with a seemingly insurmountable challenge. This led to his descent into second-dart syndrome. He regularly berated himself for being a "weak" and "stupid" person. I have seen this occur in almost every client I have worked with.

If you find yourself in this position, think of people who are suffering and put yourself in their shoes. This works even better if you personally know these people or if their struggles are similar to yours. Thinking about someone else's pain will help you escape your isolation and dramatically improve your ability to move past your stress, anxiety, or fear. As you step beyond your own suffering and think about others, like Eric Greitens did during Hell Week, this will release oxytocin in your brain and fuel your climb out of the abyss. It will help you want to reach out, connect with, and get some support from people around you as well. Perhaps more importantly for our collective salvation, this will give you the courage to use your suffering to serve others.

Psychology professor Erwin Staub calls this "altruism born of suffering." Through his research on events such as natural disasters, war, and terrorist attacks, he found people who have gone through great adversity themselves are more likely to step up and help those who need it. As an added benefit to you, helping others will improve your own health. It might even save your life.

The University of Buffalo conducted two long-term studies to track the impact serving other people has on negative or traumatic life events. They found that for those who did not regularly help others, every one of these events led to a higher risk of major health problems and a 30 percent increase in the risk of dying. For those who did regularly contribute their time and energy in service of someone beyond themselves, these events had no impact on their lives.

Love improves your health, your life expectancy, and your ability to deal with fear. Equally as powerful is fear's ability to inspire love. Collectively, the two seemingly contradictory forces enable us to achieve more than we ever thought possible. But when we take it a step higher, the real potential of Fearvana is to change the world by uniting us with each other and our planet. As Bobby Maximus says, "a family that suffers together, stays together."

My Invitation to You

"To live is the rarest thing in the world. Most people exist, that is all."
—**Oscar Wilde**

The entire journey of Fearvana can transform your life and the world around you, but all of it means nothing unless you make a choice. You can use the material in this book to better your life, improve the lives of others, do nothing, or even justify acts of inhumanity. The choice you make will provide you with your greatest education. True wisdom will emerge only when you take action and delve into the spirituality of the human soul, for that is what you will ultimately discover on your journey into Fearvana and beyond.

As you walk your path of Fearvana, remember it is about progress, not perfection. The value in the experience of Fearvana lies in the never-ending journey through fear, not the destination. As Sir Edmund Hillary said, "It is not the mountain we conquer, but ourselves."

Sometimes when you find yourself in the midst of fear, science and psychology won't help. Sometimes, as Charu does, you just need to have faith. When you set foot on your fateful frontiers to unleash the warrior within, you will need an indestructible faith in yourself, in humanity, in our collective abilities, in God— whatever it may be for you. Sometimes you will find faith is all it takes for you to step off that edge and find your wings. Love, faith, and fear can coexist. I believe they need to.

Charu's faith through suffering has been one of my greatest teachers. I hope with the lessons I have learned from her, and all the other ones I have shared with you in this book, you now have the means to explore new boundaries, shatter your preconceived limits, and keep finding your true, awakened self. Everything you do from this moment on either gets you one step closer or one step further away from the person you want to be.

This book has provided you with insight and knowledge. Use it to take action. The real lessons are learned when you take that leap into the unknown. I will leave you now with the words of renowned psychiatrist Carl Jung:

> Anyone who wants to know the human psyche will learn next to nothing from experimental psychology. He would be better advised to abandon exact science, put away his scholar's gown, bid farewell to his study, and wander with human heart through the world. There, in the horrors of prisons, lunatic asylums and hospitals, in drab suburban pubs, in brothels and gambling-hells, in the salons of the elegant, the stock exchanges, socialist meetings, churches, revivalist gatherings and ecstatic sects, through love and hate, through the experience of passion in every form in his own body, he would reap richer stores of knowledge than textbooks a foot thick could give him, and he will know how to doctor the sick with a real knowledge of the human soul.

I invite you now to put this book down, get out into the world, and begin your real education. Have fun; rise into your fears; and wander.

Acknowledgments

Writing a book has been one of the most challenging experiences of my life. There are a countless number of people who helped me become the person I am to face that challenge and ultimately bring this book into your hands.

First, I want to thank my parents. No words can express how grateful I am for the life you both have provided for me, in every way possible, and also for putting up with a very difficult child. I especially want to thank my wife as well; being married to you has been one of the greatest experiences of my life. You also created the word Fearvana, so this book wouldn't exist without you. To my brother, thank you for always caring for me and looking out for me.

To my loving grandparents, who have been a tremendous source of inspiration. I am so grateful to all of you for giving me an amazing life and teaching me how to be a better person. To the rest of my family, I thank you all for being on this amazing journey with me and am grateful for all your love and support.

Thank you to all my friends for being in my corner through all the ups and downs. Y'all are my family; I love you all. Ryan, thank you for bringing Fearvana into the online world.

To Gail, whose constant support, love, and encouragement have been a tremendous gift—thank you so much for everything. Christa, thank you for your partnership and brilliance in creating the Fearvana Foundation and for your constant, unconditional help with everything. John, Esther, and Paul, I am honored to call you my family. Thank you for helping me make this book a success. Thank you, Lindsay, for helping me use *Fearvana* to make a positive impact in the world, and for your constant love and support throughout the journey.

To His Holiness the Dalai Lama, and everyone in the office there, especially Mr. Kusho Ngawang Sonam, thank you for teaching me how to be a better human being and being a symbol of love, compassion, and peace for all of humanity. Thanks to my educators and mentors Mark Twight, Marshall Goldsmith, Seth Godin, Dean Karnazes, Chris and Janet Attwood, Rich Roll, David Rowan, Chris Guillebeau, Sebastian Copeland, Dr. Deborah Sandella, Derek Sivers, Dr. Jagdish Sheth, Marie Forleo, Keith Ferrazzi, Daniel Pink, Derek Halpern, Jaimal Yogis, Libby Gill, Justin Constantine, Kiran Mazumdar-Shaw, Sir Richard Branson, Marianne Williamson, Noah Galloway, Brene Brown, Ramit Sethi, Cal Newport, Frank Kern, Eben Pagan, Dr. Rick Hanson, Dr. Daniel Amen, and anyone whose training I have ever been a part of or whose book I have read. I am so grateful to have come across you and your work. I owe any measure of success I have achieved to the wisdom and knowledge you have imparted upon me.

I especially want to thank Steve Olsher, Marci Shimoff, Joel Roberts, and Heidi Roberts for all the one-on-one time you spent with me to help make this book a success. I wouldn't have felt confident enough to publish this without your help. Bobby Maximus, you taught me how to suffer with a smile. I am eternally grateful to you for teaching me how to constantly become better than the person I was yesterday.

Jack Canfield and everyone at the Canfield family, I am forever in your debt. You have contributed more to this book and to my life than you will ever know. I wouldn't be here without all of you.

To the team at INK Talks in India, thank you for helping me propel Fearvana into the limelight. I am forever grateful to the entire INK family for your love, support, and encouragement.

Thank you to the entire team at Morgan James Publishing for being incredibly patient with me and supporting me throughout the series of challenges I experienced in the writing of this book. I could not have asked for a better team of publishers. To my editor Amanda and the entire team at SplitSeed, thank you for your patience as well, and for helping me transform my life's work into something coherent and meaningful.

To all the people I interviewed in this book—you helped make *Fearvana* what it is. Thank you so much for that. Dale, Alice, and Charu, I am especially grateful not only for the contributions you all made to this book, but for teaching me the true meaning of courage.

To the wealth of neuroscientists, psychologists, and researchers whose work I spent countless hours on—I could never thank you enough. You all helped bring me back from the brink of suicide. This book is as much yours as it is mine.

To everyone else whose story has been included in this book, although I may not know you personally, I thank you for inspiring me and teaching me what it takes to achieve greatness. To all my brothers I served with in the US Marines, thank you for showing me what the warrior spirit really looks like. Serving with all of you has been one of the greatest honors of my life.

I almost left this section out of the book altogether because I was afraid of leaving someone out. I probably have left someone out, so to anyone I might have forgotten, I am deeply sorry. It is by no means intentional. I am grateful to everyone that has been on this beautiful journey of life with me. You have all helped mold me into the person I am today and have all helped make this book a reality.

To everyone on this list and anyone else I might have forgotten, I truly love you all and am forever grateful to each of you.

About the Author

Akshay Nanavati is a Marine Corps veteran, adventurer, and entrepreneur.

Upon overcoming drug addiction in high school, Akshay enlisted in the United States Marines, despite two doctors telling him that boot camp would kill him because of a blood disorder he was born with.

He has since run ultramarathons, climbed mountains in the Himalayas, and skied 350 miles across the world's second largest ice cap. He also spent seven months in Iraq, where his job was to walk in front of convoys to find improvised explosive devices.

After the war, he was diagnosed with PTSD and struggled with alcohol until he reached the brink of suicide. To heal his brain, he spent years studying neuroscience, psychology, and spirituality. This led to the creation of *Fearvana*.

His Holiness the Dalai Lama says "*Fearvana* inspires us to look beyond our own agonizing experiences and find the positive side of our lives." Jack Canfield, co-author of the #1 *New York Times* bestselling *Chicken Soup for the Soul* series and *The Success Principles,* says, "*Fearvana* is, without a doubt, one of the few books that really does stand out as a must-read book." Seth Godin calls *Fearvana* "counterintuitive, practical and potentially life-changing."

Today, Akshay runs a global business helping people live limitless lifestyles. His work helps fund his nonprofit, the Fearvana Foundation.

Akshay has been featured in media outlets such as *Forbes*, *Psychology Today*, entrepreneur.com, *Fast Company*, CNN, *Runners World*, NBC, Fox, and *Military Times*, among many others.

Morgan James
Speakers Group

↗ www.TheMorganJamesSpeakersGroup.com

We connect Morgan James published
authors with live and online events
and audiences whom will benefit
from their expertise.